Boris Hessen and Philosophy

Reframing the Boundaries: Thinking the Political

Series editors: Alison Assiter and Evert van der Zweerde

This series aims to mine the rich resources of philosophers in the "continental" tradition for their contributions to thinking the political. It fills a gap in the literature by suggesting that the work of a wider range of philosophers than those normally associated with this sphere of work can be of relevance to the political.

Titles in the Series

Kierkegaard and the Matter of Philosophy, Michael O'Neill Burns
Arendt, Levinas and a Politics of Relationality, Anya Topolski
The Risk of Freedom: Ethics, Phenomenology and Politics in Jan Patocka, Francesco Tava, translated by Jane Ledlie
Nietzsche's Death of God and Italian Philosophy, Emilio Carlo Corriero, translated by Vanessa Di Stefano
Lotman's Cultural Semiotics and the Political, Andrey Makarychev and Alexandra Yatsyk
Axel Honneth: Reconceiving Social Philosophy, Dagmar Wilhelm
Sartre, Imagination and Dialectical Reason: Creating Society as a Work of Art, Austin Hayden Smidt
Ontologies of Sex: Philosophy in Sexual Politics, Zeynep Direk
Reconsidering Cosmopolitanism and Forgiveness: Arendt, Derrida and "Care for the World," Christopher Peys
Transforming Politics with Merleau-Ponty: Thinking beyond the State, edited by Jérôme Melançon
Derrida and Foucault: Philosophy, Politics, and Polemics, Paul Rekret
Foucault and Governmentality: Living to Work in the Age of Control, Benda Hofmeyr
Boris Hessen and Philosophy: The Socioeconomic Roots of Classical and Modern Physics, Sean Winkler

Boris Hessen and Philosophy

The Socioeconomic Roots of Classical and Modern Physics

Sean Winkler

ROWMAN & LITTLEFIELD
Lanham • Boulder • New York • London

Published by Rowman & Littlefield
An imprint of The Rowman & Littlefield Publishing Group, Inc.
4501 Forbes Boulevard, Suite 200, Lanham, Maryland 20706
www.rowman.com

86-90 Paul Street, London EC2A 4NE

British Library Cataloguing in Publication Information Available

Library of Congress Cataloging-in-Publication Data Available

ISBN 978-1-5381-4758-0 (cloth); ISBN 978-1-5381-4759-7 (ebook)

Contents

Acknowledgments

This book is the culmination of some four years of research, and it would not have been possible without the guidance and support of so many wonderful colleagues, friends, and family. First and foremost, I would like to thank the Higher School of Economics, the Vienna Circle Institute, and Ca' Foscari University of Venice for sponsoring this research at its various phases of progress. From the Higher School of Economics, I especially extend my gratitude to Silver Bronzo, Ulrika Carlsson, Daria Drozdova, Alexei Glukhov, Andrew Haas, Stefan Hessbruggen, Tatiana Levina, Tatyana Lifintseva, Brian McLoone, Irina Sirotkina, and Aaron Wendland for your friendship and collegiality, as well as your many wonderful comments and questions at our conferences and colloquia. From the Vienna Circle Institute, I am grateful to Sabine Koch, Martin Kusch, Alexander Linsbichler, Esther Ramharter, Friedrich Stadler, and Iulian Toader for all of your generous feedback and time. And from Ca' Foscari University of Venice, my thanks particularly go out to Sascha Freyberg, Gerardo Ienna, Pietro Omodeo, and Charles Wolfe for your many helpful comments on my research and your own stimulating findings on the work of Boris Hessen. I must also extend a special thank you to Anatoliy Chusov, Steffen Ducheyne, Hiroshi Ichikawa, Koji Kanayama, Alexei Kojevnikov, Sergei Korsakov, and Maarten Van Dyck for your abundant kindness and generosity in helping to facilitate this project. Likewise, to the participants of the "Concepts and Method in Philosophy and History of Science" at Ghent University and of the "26th International Congress of History of Science and Technology" at Charles University, I am immensely grateful. Thank you also to my dearest friend, Sean O'Dubhghaill for your impeccable editing of the final version of this text. This book simply would not have been possible without help from each and every one of you.

Next, to Brill, Edizioni Ca' Foscari, *Historical Materialism*, *Science in Context*, *Society and Politics* and *Telos*, my thanks to you for allowing me to reprint in this volume the following articles in modified form:

Winkler, S. Introduction to "Boris Hessen and the Dialectics of Natural Science." *Society and Politics – Boris Hessen and the Dialectics of Natural Science* 13.1 (2019): 5–9. (See Introduction).

———. 2020. "Practice and Ideology in Boris Hessen's 'The Social and Economic Roots of Newton's *Principia*.'" *Telos* 190: 29–51. (See Introduction and chapter 1).

———. 2020. "The Materialist Dialectic in Boris Hessen's Newton Papers (1927 and 1931)." *Historical Materialism* 28.4: 202–34. (See chapter 2).

———. 2021. "Individuals and Collectives in The Philosophy of Boris Hessen: An Introduction." *Science in Context* 34.1: 121–36. (See Introduction, chapter 2, and chapter 3).

———. 2022. "A Pantheon of Great Ideas: Boris Hessen and the History and Philosophy of Science." In *Manuscripts and Documents on the History of Physics: A Historical Materialist Textbook*. Edited by Pietro Omodeo and Sean Winkler, 35–73. Venice: Verum Factum. (See chapter 3, chapter 4, and Conclusion).

Finally, this work would have no meaning without the profound love and support of my cherished friends, Marc Bennett, James DiFrisco, Patrick Eldridge, Joel Hubick, Luke McInnis, Sean O'Dubhghaill, Robyn Penney, Leendert Solotki, and Cody Staton—to the good times of the past and to so many more in the years ahead. To my Grandmother, Jackie, and to my grandparents who have passed on, Rubin, Carol and Eugene, my Uncles George, Mike, Neal, and Dean, my Aunts Connie, Teresa, and Debbie, and my cousins, Janelle, Chris, Michelle, Brian, Priscilla, and Ashley: It is truly a gift to call you all my family. To my parents, Lynn and Blake, I am eternally grateful to be your son as well as for your boundless love, generosity, and support. And finally, to my wife Olga, thank you for you for a love that knows no limits and for making each day a new adventure.

Introduction[1]

This book is meant to provide an introduction to the Soviet philosopher Boris Hessen's approach to the history and philosophy of science. Hessen is best known for delivering a talk at the Second International Congress of the History of Science and Technology in London in 1931, titled "The Social and Economic Roots of Newton's *Principia*" (hereafter, the "1931 Newton paper").[2] The paper is considered to be among the most famous in the history and philosophy of science and yet, studies of Hessen's works have largely stalled due to a number of circumstances both intrinsic and extrinsic to his oeuvre. For instance, while the 1931 Newton paper was singlehandedly his most influential work, scholars often see it as the least typical of his writings.[3] And despite its impact in North American and Western European contexts in precipitating numerous subfields in the history and philosophy of science (i.e., the Sociology of Science, Science and Technology Studies, the Visible College, etc.), save for a few notable exceptions, this influence was largely indirect, leaving direct study of Hessen's thought largely at a standstill there.[4] Hessen also received hardly any attention in either the former Soviet Union or Eastern Bloc.[5] However, as the old adage goes, "what's old becomes new again" and the past two decades have witnessed something of a minor renaissance in Hessen studies. There are ever more translations of his 1931 Newton paper, with it now being available not only in English and Russian but also in French, German, Greek, Korean, Italian, Japanese, Spanish, and Swedish as well. Once seen as the paradigmatic example of the so-called "externalist" approach to the history and philosophy of science, in revisiting Hessen's thought, contemporary scholars have come to realize that it offers a far more complex and nuanced perspective than previously thought. Moreover, scholars have become less content with studying the 1931 Newton paper alone and have begun delving deeper into his oeuvre, along with issuing translations of his other works for the very first time. While some claim that the lack of attention to his other works had been due to their being rather conventional contributions to the field, the past years of study reveal this to be far from the case.[6]

So, who exactly was Boris Hessen?[7] Born in August 1893, he was the child of Jewish parents in Elisavetgrad, which was part of the Russian Empire at that time and which is today known as Kropyvnytskyi, Ukraine.[8] He studied physics and mathematics at the University of Edinburgh, given that he had to seek education outside of the empire because of quotas on the number of Jewish students allowed into the Russian university system.[9] He became committed to Marxist ideals as a young man, and when the October Revolution and subsequent Russian Civil War came, he joined the Red Army and the Communist Party, specifically playing an important role in educating workers in mathematics and the natural sciences in his hometown. He was a true believer in the importance of bringing scientific education to the people to whom it had been previously denied.[10] In the aftermath of the Civil War, he would go on to graduate from the Institute of Red Professors, a graduate institute in which students received an education in Marxist theory, and later went on to become the director of physics at Moscow State University.[11] As a historian and philosopher of science, Hessen belonged to a group of theoreticians known as the Deborinites (so-called after Abram M. Deborin[12]) or "dialecticians," who took a more speculative approach to Marxist dialectical materialism, and saw the true potential of such a theory in its receptivity to non-Marxist philosophical and scientific positions.[13] As a Deborinite, Hessen focused heavily on analyzing how contemporary scientific theories, such as quantum mechanics and relativity theory, could be reconciled with this theory.[14] In 1931, he was selected to be a member of the Soviet contingent who were to attend the aforementioned London conference; the circumstances surrounding the decision to send him (and to send a contingent at all) remain something of a mystery to this day.[15] Hessen wrote what would become his most famous paper for this conference, the 1931 Newton paper, which offered a dialectical materialist reading of the emergence of Newtonian physics.[16] Once considered "one of the most influential reports ever presented at a meeting of historians of science,"[17] it caused a sensation among notable thinkers such as J. D. Bernal, J. B. S. Haldane, Lancelot Hogben, Hyman Levy, Robert K. Merton, and Joseph Needham—some of whom were even in attendance at the London conference—and it continues to exert a subterranean influence on the subject to this day.[18] On August 21, 1936, Hessen was arrested by the NKVD on the dubious charges of "involvement in the terrorist activities of a Trotskyist-Zinovievist conspiracy," part of which involved "complicity in the 1934 murder of S. M. Kirov and of planning terror attacks on leading officials."[19] He was brought to trial on December 20, 1936, and executed by firing squad the same day, making him one of the first among what would become the countless victims of Joseph Stalin's regime in the so-called "Great Purge." He simply vanished without a trace to his family, friends, and colleagues. Nothing of his apprehension and execution was made public until after his rehabilitation in 1955; many of Hessen's biographical details continue to be unearthed.[20]

How is it, then, that this nearly forgotten figure is only now beginning to attract more direct scrutiny, after almost a century? One might say that the mounting global, medical, ecological, political, socioeconomic, and cultural disturbances of the present suggest the growing need to imagine new ways of being and thinking on a grand scale. Perhaps, we have reached a point of having to, once again, think in terms of grand narratives. Hessen, a theoretical physicist, historian, philosopher of science, and a Marxist, promoted his own kind of grand narrative, and while he was a major intellectual figure in the early Soviet period and a proponent of dialectical materialism, he was not a proponent of it in its official, state doctrinal form known as "diamat." Instead, Hessen was considerably more open to theories that were not consistent with this ideology.[21] One might say, then, that the sources of the conflict that had once stood as obstacles in the way of Hessen studies, in both the East and West, are now the very thing that are propelling interest in it forward—a way of studying an underexplored path from the past to see if it can, in any way, shine a light on a better way forward.

What grand narrative does Hessen advance exactly? As indicated, he may have been a dialectical materialist, but he also insisted that it be subject to ongoing developments in order to both accommodate and facilitate new discoveries in the natural sciences. Moreover, he encouraged collaboration between scientists and the working class in order to encourage the cross-pollination of theoretical and practical knowledge. However, even this overview is far from exhaustive, given the full range of thought that Hessen grappled with and the overwhelming number of subjects across his oeuvre: classical mechanics versus quantum mechanics and relativity theory, dynamic versus statistical laws, reversible versus irreversible natural processes, macrocosmic versus the microcosmic natural processes, social relations/means of production versus natural science, and so on. Many have assumed that the core of his thought had been expounded in the 1931 Newton paper, but this is hardly his magnum opus, given that his ideas span the course of several books, correspondences, numerous articles, and encyclopedias entries as well as lectures. He also engages with the works of a nearly endless cast of figures, both historical (Sir Francis Bacon, René Descartes, G. W. F. Hegel, Karl Marx, Sir Isaac Newton, etc.) and contemporaneous (Albert Einstein,

Ludwig Boltzmann, Werner Heisenberg, V. I. Lenin, James Clerk Maxwell, Erwin Schrödinger, Marian Smoluchowski, etc.). There is, then, still much work to be done in order to understand if and how Hessen's various works all fit together and what, if anything, they might have to tell us about the present day.

A summation of Hessen's thought certainly far exceeds the capacities of any single volume. This book, then, in no way pretends to be a comprehensive treatise but, rather, sets out to introduce Hessen's thought via a series of interrelated reflections that shed light on his foundational intuitions. The guiding thesis is that Hessen offers one of the most concerted efforts to construct a serious dialectical materialist philosophy, one which would be capable of, on one hand, grasping the complexities of constructing a new society and, on the other hand, helping to charter a new way of understanding scientific discovery. I develop this thesis over the course of four chapters. In chapter 1, Practice and Ideology, I provide an analysis of the two foundational concepts for Hessen's dialectical materialist approach to the history and philosophy of science; namely, practice and ideology, paying special attention to his 1931 Newton paper. Chapter 2, Idealism and Materialism, examines Hessen's belief in the importance of ideological conflict for the emergence of new natural scientific theories, particularly in the dialectic between the philosophical persuasions of idealism and materialism; this is a position that I analyze by focusing especially on his 1931 Newton paper alongside his contributions to the Soviet journal, *Under the Banner of Marxism*. In chapter 3, Modern Physics, I explain Hessen's efforts to read quantum mechanics and relativity theory in light of dialectical materialism, while focusing especially on his Preface to the Russian translation of Arthur Haas' *Wave Mechanics and the New Quantum Theory*, *The Main Ideas of Relativity Theory* and contributions to the *Soviet Encyclopedia*. Finally, in chapter 4, A Pantheon of Great Ideas, I provide a summary of Hessen's thought by demonstrating how what had appeared to be disparate facets of his thinking actually come together in the context of his *Manuscripts and Documents on the History of Physics*. It is my hope that this book will facilitate a better understanding of the strengths and weaknesses of Hessen's work, along with precipitating the continued evaluation of his legacy in the context of debates within the history and philosophy of science.

NOTES

1. This introduction is a revised version of my Introduction, "Boris Hessen and the Dialectics of Natural Science," *Society and Politics – Boris Hessen and the Dialectics of Natural Science* 13.1 (2019): 5–9, and also consists of passages from my "Practice and Ideology in Boris Hessen's 'The Social and Economic Roots of Newton's

Principia'," *Telos* 190 (2020): 29–51 along with my "Individuals and Collectives in the Philosophy of Boris Hessen," *Science in Context* 34.1 (2021): 121–36.

2. B. Hessen, "The Social and Economic Roots of Newton's *Principia*," in *The Social and Economic Roots of the Scientific Revolution: Texts by Boris Hessen and Henryk Grossmann*, ed. G. Freudenthal and P. McLaughlin and trans. P. Schmrat (Dordrecht: Springer, 2009 [1931]), 41–101.

3. L. R. Graham, "The Socio-Political Roots of Boris Hessen: Soviet Marxism and the History of Science," *Social Studies of Science* 15 (1985): 705, 707; W. Schäfer, "Boris Hessen and the Politics of the Sociology of Science," *Thesis Eleven* 21 (1988): 113, 114.

4. See G. Ienna and G. Rispoli, "Boris Hessen at the Crossroads of Science and Ideology: From International Circulation to the Soviet Context," *Society and Politics – Boris Hessen and the Dialectics of Natural Science* 13.1 (2019): 37–64.

5. Schäfer, "Boris Hessen and the Politics of the Sociology of Science," 103.

6. Graham, "The Socio-Political Roots of Boris Hessen," 707; D. Joravsky, *Soviet Marxism and Natural Science, 1917–1932* (London: Routledge, 2009), 286; Schäfer, "Boris Hessen and the Politics of the Sociology of Science," 114.

7. For more on Hessen's biography see, among others, G. Freudenthal, G. and P. McLaughlin, "Boris Hessen: In Lieu of a Biography," in *The Social and Economic Roots of the Scientific Revolution: Texts by Boris Hessen and Henryk Grossmann*, ed. G. Freudenthal and P. McLaughlin (Dordrecht: Springer, 2009), 253–55; R. L. Winkler, "Ein unveröffentlichtes Manuskript von Boris M. Hessen: 'Materialien und Dokumente zur Geschichte der Physik' (Druckfahnen – 1936 [?], ca. 700 Seiten, russisch): Kurzvortrag vor der Klasse für Sozial- und Geisteswissenschaften am 14.12.2006, gewidmet Boris Hessen anläßlich seines 70. Todestages am 20. Dezember 2006]," *Sitzungsberichte der Leibniz-Sozietät der Wissenschaften zu Berlin* 92 (2007): 133–52; С. Н. Корсаков, А.В. Козенко, and Г.Г. Грачева, Борис Михайлович Гессен. 1893–1936 (Москва: Наука, 2015).

8. Корсаков, Козенко and Грачева, Борис Михайлович Гессен, 6–7.

9. Ibid., 7.

10. Ibid., 10, 11; C. A. J. Chilvers, "The Tragedy of Comrade Hessen: Biography as Historical Discourse," in *The History and Poetics of Scientific Biography*, ed. T. Söderqvist (Burlington: Ashgate Publishing, 2007), 112.

11. Chilvers, "The Tragedy of Comrade Hessen," 112.

12. Abram M. Deborin was a Soviet philosopher and member of the Academy of Sciences of the USSR. Between 1926 and 1931, he served as the editor of the prominent Soviet journal, *Under the Banner of Marxism*, which was a major battleground for philosophical debates concerning Marxist theory. Brian Kassof, "A Book of Socialism: Stalinist Culture and the First Edition of the *Bolshaia Sovetskaia Entsiklopediia*," *Kritika* 6.1 (2005): 30.

13. Schäfer, "Boris Hessen and the Politics of the Sociology of Science," 114.

14. P. Josephson, *Physics and Politics in Revolutionary Russia* (Berkeley: University of California Press, 1991), 269; Freudenthal and McLaughlin, "Boris Hessen," 255; Joravsky, *Soviet Marxism and Natural Science*, 286.

15. P. G. Werskey, Introduction to *Science at the Cross Roads – Papers Presented to the International Congress of the History of Science and Technology, Held in London from June 29th to July 3rd, 1931 by the Delegates of the USSR*, ed. Nikolai Bukharin (London: Frank Cass and Co. Ltd., 1971), xi.

16. Hessen, "The Social and Economic Roots of Newton's *Principia*," 41–42.

17. Graham, "The Socio-Political Roots of Boris Hessen," 705.

18. Ibid., 705–6; Ienna and Rispoli, "Boris Hessen at the Crossroads of Science and Ideology," 39–49; Schäfer, "Boris Hessen and the Politics of the Sociology of Science," 106.

19. G. Gorelik and A. W. Bouis, *The World of Andrei Sakharov: A Russian Physicist's Path to Freedom* (Oxford: Oxford University Press, 2005), 59; Freudenthal and McLaughlin, "Boris Hessen," 255.

20. Gorelik and Bouis, *The World of Andrei Sakharov*, 59; Freudenthal and McLaughlin, "Boris Hessen," 255.

21. Joravsky, *Soviet Marxism and Natural Science*, 185–87; Josephson, *Physics and Politics in Revolutionary Russia*, 246; H. Sheehan, *Marxism and the Philosophy of Science: A Critical History* (Atlantic Highlands: Humanities Press International, Inc., 1993 [1985]), 179–82.å

Chapter One

Practice and Ideology[1]

In this first chapter, I intend to offer an interpretation of two central, albeit still relatively little understood, concepts in Boris Hessen's philosophical system: practice and ideology. Hessen famously enumerated four claims (hereafter, the Hessen theses) about the social and economic roots of one of the greatest physicist of all time, Sir Isaac Newton's natural philosophy in his "The Social and Economic Roots of Newton's *Principia*" (the 1931 Newton paper). The Hessen theses can be paraphrased as follows: First, all of the theoretical content in Newton's natural philosophy pertaining to physics is traceable to practical problems related to the development of the machine technology of the day; second, all of the theoretical content that pertains to metaphysics in Newton's natural philosophy is traceable to his theology, which is informed by his class standpoint; third, the theoretical content of Newton's physics did not go beyond practical problems related to the development of the machine technology of the day; and fourth, the loss of workers' occupations, through the substitution of machines for human labor, could not be overcome through the wrecking of machines but from workers seizing the means of production and by eliminating class-based society.[2]

Underlying his four theses, Hessen makes a basic assumption about the meaning of and relationship between practice and ideology; terms that are essential to his entire position, yet which are not defined anywhere in the 1931 Newton paper. Rather, he prefers to refer to these notions through examples, rather than in generalizations, and provides neither any concrete definitions nor any summative account of their relationship to one another. Moreover, Hessen derives these notions from the thought of Karl Marx, Friedrich Engels, and Vladimir Lenin, whose ideas he often treats as monolithic, even though there are, arguably, substantial differences between them as well as substantial differences between the individual texts found within each of their oeuvres. While previous studies of Hessen's 1931 Newton paper have made significant progress in renewing interest in this once-forgotten work, the lack of attention paid to this distinction between practice and ideology precludes our ability to have a clear sense of how the theses all hang together.[3]

In this chapter, then, I propose the following position: For Hessen, practice can be defined as the transformation of things in-themselves into things for-us, but it also involves the transformation of things in-themselves into things for-themselves; for him, ideology can be defined as the specific difference between practice and theory, when the practical roots of theory are concealed. In the section, "Rehabilitating" Hessen, I explain how the Hessen theses have once again garnered attention and further explain how a better understanding of Hessen's work requires an appreciation of his Deborinite conception of dialectical materialism. In the next section, The Social and Economic Roots of Newton's *Principia*, I explain the Hessen theses further and identify means and relations of production as the two kinds of practice presented throughout the Newton paper. In Dialectical Materialism(s), the third section, I enumerate the conceptions of practice employed by Marx, Engels, and Lenin and how Hessen, as a Deborinite, integrated aspects of these conceptions into a single view in the Newton paper. Finally, in the last section, Bukharin's "Praxeology," I show that Nikolai Bukharin's "Theory and Practice from the Standpoint of Dialectical Materialism" (hereafter, "Theory and Practice") provides a proper companion piece to the 1931 Newton paper, given that Bukharin's account of practice and theory opens up the possibility of better understanding Hessen's account of practice and ideology.[4] To advance my argument, I draw from the 1931 Newton paper, the writings of Marx, Engels, Lenin, and Bukharin as well as the secondary literature.

"REHABILITATING" HESSEN

In this first section, I intend to show that prior to undertaking an examination of practice and ideology in Hessen's 1931 Newton paper, one must first have a proper sense of how to approach Hessen as a theoretician. For decades, Hessen and his work had been caricatured and misrepresented far more than they had been carefully studied. In one instance, his 1931 Newton paper was even portrayed as a "narrow-minded piece of bigoted dogmatism at its Stalinist crudest."[5] Luckily, this reductive tendency has begun to fall out of favor in the last decade or two, thanks in no small part to Gideon Freudenthal's 2005 article, "The Hessen-Grossmann Thesis: An Attempt at Rehabilitation," as well as his 2011 article with Peter McLaughlin, "Classical Marxist Historiography of Science," both of which argue that in the aftermath of the end of the

Cold War, the time had finally come to reread the Hessen theses with a fresh pair of eyes in order to extricate them from this association with Stalinism.[6] In this section, then, I will explore how Freudenthal and McLaughlin took the important first step of reinvigorating the study of Hessen's work, before proceeding to claim that the next step should consist of understanding how Hessen's work is the epitome of a Deborinite approach to dialectical materialist philosophy.

Freudenthal and McLaughlin's works have done a great service in asserting that the Hessen theses deserve to be revisited as quintessential examples of a dialectical materialist approach to the history and philosophy of science. To distance the Hessen theses from their past vulgarizations, these authors advance the, what they call, "Hessen-Grossmann thesis," which condenses certain elements from the 1931 Newton paper with the support of by the work of Henryk Grossmann. Grossmann was a twentieth-century economist, historian, and socialist working in Western Europe and was member of the Institute for Social Research, the home of the "Frankfurt School." He composed his contributions to the history of early modern mechanistic philosophy—"The Social Foundations of the Mechanistic Philosophy and Manufacture," "Descartes and the Social Origins of the Mechanistic Concept of the World," a letter to Friedrich Pollock and Max Horkheimer from August 23, 1935, and a review of G. N. Clark's *Science and Social Welfare in the Age of Newton* and George Sarton's *The History of Science and the New Humanism*—independently of Hessen; however, Grossmann agrees with Hessen in many respects and, in one case, even mentions him by name.[7] Freudenthal and McLaughlin hope to present Hessen's ideas as an effort to "conceptualize science as one kind of labor within the system of social production" through the proposal of the Hessen-Grossmann thesis;[8] that is, they hope that people will treat it as one of the "classic programmatic examples of Marxist historiography of science."[9] In advancing this position, Freudenthal and McLaughlin have successfully contributed to taking a socioeconomic understanding of the history and philosophy of science seriously and to opening the door once more to a renewed study of Hessen's work.

To fully grasp the whole of Hessen's vision, I maintain that the next step in the study of his oeuvre should consist of examining his dual premises of practice and ideology; however, taking this step requires an appreciation of his background as a Deborinite dialectical materialist. The meaning and significance of dialectical materialism was in considerable flux in the Soviet Union of the late 1920s, as epitomized in the debate between two prominent camps of thinkers: the mechanists and the Deborinites or dialecticians. For the former, dialectical materialism was a reductionist natural philosophy that resembled Newtonian mechanics (minus the theological elements), and this camp was exceedingly hostile to any nonmaterialist positions.[10] For the latter, dialectical materialism was an antireductionist natural philosophy that continually evolved, and this camp saw it as necessary to entertain a more speculative approach to nonmaterialist positions.[11] The Deborinites favored the incorporation of Hegelian philosophy into dialectical materialism, predicated in no small part by Lenin's initiative to encourage the study of G. W. F. Hegel's philosophy for the sake of developing heuristic accounts of materialist dialectics.[12] The Deborinites eventually won out over the mechanists, and even enjoyed a brief period of prominence, but this was quickly cut short after Joseph Stalin's consolidation of power in 1928 and the "Great Break."[13] Stalin insisted on the Great Break because of the failure of other socialist revolutions to metastasize worldwide, thereby isolating the Soviet Union, along with the country's pressing need to modernize. To this end, he replaced Lenin's New Economic Policy (NEP) with heavy industrialization and the collectivization of agriculture, while also insisting on the transformation of what had for a brief time been a relatively liberal academic sphere into a "proletarianized" one; that is, one directed toward the pursuit of knowledge for the sole purposes of economic and technological development. In response to the scale and gravity of this event, Hessen and the Deborinites came under fire as their more speculative efforts were seen as counterproductive to the construction of socialism.[14] The Deborinites's change of fortune is perhaps epitomized in an article that was coauthored by Mark Mitin, Vasiliy Raltsevich, and Pavel Yudin and titled "On the New Tasks of Marxist-Leninist Philosophy."[15] The Deborinites fought back as exemplified in an article coauthored with Hessen, among several others, titled "On the Fight on Two Fronts," wherein they attest to the importance of theoretical speculation for the purposes of building socialism. As they write,

> [i]n the field of philosophy is the knot in which all the most important questions of the *theory and practice of the struggle of the proletariat in our era* are connected. . . . The problem of time and space in relativity theory, the problems of the ether, waves and particles in physics, the problem of heredity and variability in biology, the problems of cosmogony, geology, the history of the elements, physiology and the study of the higher nervous activities of the fundamentals of medicine necessarily lead modern science to philosophical questions, to the doctrine of causality, the problem of the relation of matter and motion, continuity and discontinuity, chance and necessity. On one hand, they provide tremendous material for developing the theory of materialist dialectics, and on the other hand, they can be studied and understood only on its basis.[16]

The effort was to little avail, however, and the definitive blow to the Deborinites came on January 25, 1931, when Stalin decreed that his own approach to dialectical materialism was to be the official ideology of the Communist Party. He would later codify the basic tenets of this doctrine in his *Dialectical and Historical Materialism*.[17] Despite the shifting political terrain of his day, though, Hessen's thought remained emblematic of the Deborinite approach to dialectical materialism from beginning to end; that is, he saw it as a philosophy that mediated socialist construction alongside the relative limits of socioeconomic development, technological progress and the absolute limits of the laws of nature, and as something which needed to remain open to its own theoretical development. This can be seen in the aforementioned article, which states that "[t]he struggle for the philosophical materialism of Marx, i.e. for *dialectical materialism*, is the necessary link in the struggle for communism. . . . In the field of philosophy, it is that knot in which all the most important questions of the *theory and practice of the struggle of the proletariat in our era* are connected."[18] The task of developing a possible account of practice and ideology in the 1931 Newton paper, then, falls to studying Hessen's work in light of his Deborinite background.

THE SOCIAL AND ECONOMIC ROOTS OF NEWTON'S *PRINCIPIA*

In this section, I intend to walk the reader through the actual Hessen theses themselves, so as to lay the groundwork for our ability to identify their specifically Deborinite features. Hessen's 1931 text is a reading of Newton's *Mathematical Principles of Natural Philosophy* (hereafter, *Principia*), a volume that requires almost no introduction. It is a work of staggering ambition and not only does it epitomize the so-called "mechanistic natural philosophy" of the seventeenth century, but it also continues to be relevant for philosophy and for the natural sciences today, more than three centuries after its composition. Generally speaking, the achievements of the *Principia* are attributed to Newton's personal genius. For instance, Alexander Pope writes in his "Epitaph on Sir Isaac Newton," "Nature and nature's laws lay hid in night; God said 'Let Newton be!' and all was light,"[19] and Alfred North Whitehead speculates in his "The First Physical Synthesis," "[j]ust think what the course of human history would have been if these two men had not appeared in the world."[20]

In his Newton paper, however, Hessen maintains that thinkers such as Pope and Whitehead ignore an important aspect of the *Principia* that cannot be understood by reading this text alone, namely, the "complete overlap between the topics that concerned physics in that era, which arose out of economic and technical needs, and the contents of the *Principia*, which constitute in the full sense of the word an outline and systematic solution of the entire range of the main physical problems."[21] For Hessen, this is not only true of Newton's *Principia*, but of early modern, natural philosophy as a whole. He writes that "[t]he dazzling flowering of natural science during the 16th and 17th centuries resulted from the disintegration of the feudal economy, the development of merchant capital, of international maritime relations and of heavy (mining and metallurgical) industry."[22]

Paraphrasing Lenin's *Karl Marx*, Hessen maintains that the oversights of thinkers like Pope and Whitehead stem from "two chief defects"[23] in their approach to history. As he writes, the first defect is that "[e]arlier historical theories examined only the *ideological* motives in the historical activities of human beings," while "[t]he second defect . . . is the view that the subject of history is not the mass of the people, but individuals of genius."[24] To overcome these two shortcomings, Hessen aims "to apply the method of dialectical materialism and Marx's conception of the historical process to an analysis of the genesis and development of Newton's work within the context of the period in which he lived and worked;"[25] that is, he aims to read Newton's *Principia* through the lens of Marx's dialectical materialism. The underlying assumption of this approach is that "[p]ractice should not be explained by ideas, but on the contrary, ideological structures should be explained by material practice."[26] As indicated previously, unfortunately, Hessen does not provide his own definitions or explanations of the relationship between these two terms or even whether some or all of Newton's natural philosophy is ideological; thus, one is left to extract some meaning from his theses and references to other thinkers. Hessen's four theses proceed as follows.

In thesis 1, "The Physical Themes of the Era and the Contents of the *Principia*," Hessen begins by pinpointing the exact historical context in which Newton flourished, namely, late seventeenth and early eighteenth centuries in England. Economically speaking, this context was characterized by "the rule of privative property"[27] as England was undergoing the unraveling of feudalism, but it had not yet arrived at full-scale, industrial capitalism.[28] The prevailing form of economic production of the day was "manufacture," or the organization of labor through the putting-out system or workshop system; these were organizational forms orchestrated by the rising merchant class and their accumulation of capital. Paraphrasing Marx's analysis of manufacture in *Capital, vol. 1*, Hessen maintains that

> [m]anufacture developed out of handicrafts in two ways. On the one hand it arose from the combination of heterogeneous independent handicrafts, which lost their independence, and on the other hand it arose from the co-operation between craftsmen in the same craft, which broke down the particular process into its component parts and led to a division of labour within manufacture.[29]

This transformation manifested as the shift from consumption-based economic production to exchange-based economic production, coinciding with not only a division of labor in the workshop but also with a division of labor in society as a whole.[30] Although a strict division of labor would not emerge until the eighteenth century, the emergence of manufacture saw a crucial split in the division of mental labor from physical labor and provided ripe conditions for the increased significance of the role played by machines in economic production.[31] The development of machines primarily took place in the spheres of communication, industry, and warfare.[32] The rise of communication occurred just as merchant capital witnessed the development of bigger and faster ships, while the rise in industry witnessed the demand for currency and the increasing role played by the merchant class in waging warfare and in the development of firearms.[33] In each instance, Hessen appears to identify three distinct moments in which practical problems came to be translated into theoretical problems: First, practical demands arose within a given industry; second, theoretical abstractions of those practical demands arose within a given industry; and third, theoretical abstractions of practical problems arose across different industries.[34] Hessen summarizes all of the practical problems in the physics of the day and the spheres of economic production from which they derive:

> the first group of problems constitutes the physical problems relating to lifting devices and transmission mechanisms that were important to the mining industry and the art of building.
>
> The second group of problems is of great importance for artillery and constitutes the main physical problems relating to ballistics.
>
> The third group of problems is of great importance to the problems of the drainage and ventilation of mines, the smelting of ore, canals and lock construction, internal ballistics and designing the shape of ships.
>
> The fourth group is of enormous importance to navigation.
>
> All these are fundamentally mechanical problems.[35]

He proceeds by showing all of the theoretical problems that derive from these practical problems:

1. The problem of simple machines, inclined planes and general problems of statics.
2. The free fall of bodies and the trajectory of a projectile.
3. The laws of hydro- and aerostatics, and atmospheric pressure. The pump, the motion of bodies through a resistant medium.
4. Problems of celestial mechanics, the theory of tides.[36]

The rising merchant class resorted to founding their own societies, so that they could patron research that was favorable to their interests and also because medieval/early modern universities were either ill-equipped or were actively unwilling, in some cases, to address such subject matter. Newton, being a member of one such society, the Royal Society of London, would have been exposed to each and every one of these practical and theoretical problems and this is corroborated by the inclusion of each and every one of the latter in his *Principia*.[37]

In thesis 2, "The Class Struggle during the English Revolution and Newton's Worldview," Hessen acknowledges that not all of the theoretical problems found in the *Principia* are traceable to practical problems with machines. For instance, at the helm of Newton's material universe stands an immaterial, providential God, but it is by no means clear that such a concept had to be bound up with the mechanistic elements of his system.[38] After all, there appear to have been not just one, but a plethora of mechanistic, natural philosophical systems during the early modern period, some of which conceived God in a different manner and some of which omitted God from the equation altogether. For René Descartes, for instance, it was necessary for God to preside over a mechanistic universe, but he "needed God only to prove that the quantity of motion in the universe remains constant."[39] For Thomas Hobbes, a mechanistic universe "came to the defense of royal power and absolutism and encouraged the repression of the people,"[40] while for Robert Overton, a mechanistic universe is "blatantly materialistic and atheistic."[41] Hessen concedes that "[i]t would be a gross oversimplification to derive *every problem* studied by various physicists, and *every task* they solved, directly from economics and technology."[42] Even though

> [t]he economic situation is the basis[,] . . . the development of theories and the individual work of a scientist are also affected by various superstructures, such as political forms of the class struggle and its results, the reflection of these battles in the minds of the participants—in political, juridical, and philosophical theories, religious beliefs and their subsequent development into dogmatic systems.[43]

Thus, while "the idealistic views of Newton are not incidental, but organically bound up with his conception of the universe,"[44] they remain unessential to a mechanistic conception of the material universe as such. The specific difference between Newton's mechanistic conception of the universe and that of the other aforementioned thinkers is, for Hessen, their respective class standpoints. Newton was a "typical representative of the rising bourgeoisie;"[45] Hessen even calls him a "typical child of the class compromise of 1688 [or, Glorious Revolution],"[46] between the aristocracy and the merchant class, himself a member of the latter. The materialism of the day in England was that of either Hobbes, who was considered royalist and absolutist, or that of Overton, who was, according to Hessen, a proto-Bolshevik, and the materialism of the day in France was bourgeois but revolutionary.[47] So, materialism was not conducive to the ideology corresponding with Newton's class, which sought democratization and religious tolerance.[48] There was a providential divine that was responsible for animating lifeless matter within the rising bourgeoisie's belief system. As Hessen maintains, "[t]his unique 'division of labour' in the government of the universe between God and causation was characteristic of the way in which the English philosophers interwove religious dogma with the materialistic principles of mechanistic causation."[49] Lastly, he maintains that absolute space and absolute time mark the pivotal points around which the relation between the theological and physical aspects of Newton's system come to be inverted, placing God, rather than the material universe, as the first principle.[50]

Thereafter, in thesis 3, "Engels' Conception of Energy and the Lack of the Law of Conservation of Energy in Newton," Hessen tries to explain why Newton only considers one type of motion, namely, mechanical motion. In Newton's time, and even today,

> [m]echanics studies the form of motion that consists in the simple displacement of bodies in space. Nevertheless, in addition to this form of motion there are a number of other forms of motion of matter, in which mechanical displacement recedes to the background by comparison with new specific forms of motion [51]

Newton simply could not have conceived of other kinds of motion, Hessen argues, because the practical problems that made such a concept possible had not yet arisen. The conversion of, for instance, thermal energy into mechanical motion was not possible until the steam engine came to be used in industrial production and only after its functions were extensively studied in thermodynamics.[52]

Thesis 4, "The Machine-Wreckers in Newton's Age and the Present-Day Wreckers of the Productive Forces," is the most commonly overlooked of the four theses. Here, Hessen refers to "machine wreckers," or to workers who took to destroying machines when they began to see automation as a threat to their employment. He argues that while this gesture expresses the correct sentiment of protesting wage-labor, it treats machines as the source of the loss of their livelihood, when the problem is wage-labor itself, which is a product of a class-based society organized around merchant capital. He writes that the destruction of machines does not amount to a genuine liberation of the workers because the system of wage-labor itself remains intact. In a sense, machine-wrecking is a sign that the working class is still dominated by the ruling class's ideology, by "machine fetishism" or by a fixation with machines as the motor of society.[53] Hessen identifies the importance of not simply organizing machine-wrecking but uniting as a class via the development of class consciousness. Hessen betrays Hegelianism in the closing arguments of his paper, making a peculiar formulation that "the proletariat does not remain a mere spectator. It is not only the object, but the subject of the process."[54] He maintains that when the proletariat assumes control over the means of production, a new science will unfold from the classless society that will follow.[55]

Many of the specific, historical details found in the 1931 Newton paper are Hessen's own contributions or citations from other historical works, but the core argument of each thesis appears to be the crystallization of a claim made previously by either Marx or Engels. For instance, thesis 1 appears to derive from Marx and Engels's *The German Ideology*,[56] a letter by Marx to Engels from September 25, 1857,[57] Engels' *Socialism: Utopian and Scientific*[58] as well as his *Dialectics of Nature*,[59] whereas the core claim of thesis 2 can be traced to Marx and Engels's *The Holy Family*,[60] Marx's "The Bourgeoisie and the Counter-Revolution,"[61] Engels's aforementioned texts,[62] as well as his letter to Schmidt from August 5, 1890.[63] Thesis 3 derives from Marx's *Capital, vols. 1*[64] and *3*[65] and thesis 4 derives from *Capital, vol. 1*[66] and Engels's *Dialectics of Nature*.[67] While it might seem that identifying these sources would solve the problem of finding an account of practice and ideology, it actually only exacerbates it, as Hessen treats *en bloc* sources that do not necessarily agree in terms of methodology. This should not be taken to mean that Hessen is orthodox in his interpretations of Marx or Engels either, insofar as these texts lend themselves to various interpretations.

Matters become no less obscure when one turns to the sources of Hessen's basic, methodological claims. He attributes his methodological claims to Marx and Engels's *The German Ideology* and Marx's Preface to *A Contribution to the Critique of Political Economy*, two works with different aims that stem from different phases of Marx and Engels's development. From these references, however, one might say for certain that Hessen agrees with them about examples of practice and ideology; "definite relations of production" and "material forces of production" (hereafter, relations of production and means of production) are examples of the former, whereas "legal, political, religious, artistic or philosophic" structures are examples of ideological structures[68] of the latter. Like the previous citations, these passages also lend themselves to several interpretations.

What these references do indicate, however, is that Hessen's overall strategy in presenting four theses. Upon closer scrutiny, theses 1 and 3 and theses 2 and 4 appear to form two pairings that run in parallel. Theses 1 and

3 establish the expanse and limit of the horizon of knowledge as engendered by machine technology, while theses 2 and 4 establish the expanse and limit of the horizon of knowledge as engendered by class standpoint. When taken together, these two pairs determine the horizon of possible objects that are available for theoretical reflection in a certain socioeconomic context. One can verify that this identifies Hessen's approach with the history of philosophy, based on his claim that "[a] historical study of forms of motion should be conducted from two perspectives. We must study the historical sequence of the forms of motion as they appear in the development of the science of physics in human society. The second perspective is to study the 'natural science of the development of matter."[69] This captures the two-fold essence of, what Hessen calls, a "Marxist classification of the sciences,"[70] studying the appearance of the forms of motion on the basis of the means of production as well as in terms of the relations of production at a given moment of history.

DIALECTICAL MATERIALISM(S)

With this third section of the chapter, I intend to explore the philosophical premises to which Hessen adhered and, likewise, to demonstrate how they reflect the Deborinite nature of his approach to dialectical materialism. When Hessen claims dialectical materialism as his method, this does give one some sense about the essential texts on the basis of which his account of practice and ideology came into existence.[71] In the 1931 Newton paper, for instance, he attributes his methodology to Marx, Engels, and Lenin and to their respective texts that contain accounts of practice and ideology, including Marx's "Theses on Feuerbach," Engels's "Ludwig Feuerbach and the End of Classical German Philosophy," and Lenin's *Materialism and Empirio-Criticism*. As a Deborinite, however, he would have had access to, as well as would have taken particular interest in, Lenin's *Philosophical Notebooks*. A closer examination of each text reveals possible conceptions of practice and ideology that could be applicable to Hessen's 1931 Newton paper.

One of the fundamental premises of any strand of dialectical materialism is the eleventh of Marx's "Theses on Feuerbach," in which he offers the startling declaration that "[p]hilosophers have hitherto only *interpreted* the world; the point is to *change* it."[72] Although Marx insinuates more than he informs here, he nevertheless challenges the connotation of philosophy as a disinterested pursuit of the truth. Truth is never disinterested and, as he states in his second thesis, "[t]he question whether objective trust can be attributed to human thinking is not a question of theory but is a *practical* question. Man must prove the truth, i.e., the reality and power, the this-worldliness of his thinking in practice. The dispute over the reality or non-reality of thinking which is isolated from practice is a purely *scholastic* question."[73] In other words, the philosopher tries to contemplate the truth of a ready-made world, but the truth of the world can only be tested through appropriating, and thereby transforming, it.

In "Ludwig Feuerbach and the End of Classical German Philosophy," Engels elaborates upon the "Theses on Feuerbach," offering both a possible definition of practice, as well as several examples by which to clarify Marx's more cryptic formulations. As Engels writes,

> [t]he most telling refutation of . . . all . . . philosophical quirks is practice, namely, experimentation and industry. If we are able to prove the correctness of our conception of a natural phenomenon by brining it about ourselves, producing it out of its conditions and making it serve our own purposes into the bargain, then the graspable Kantian 'thing-in-itself' is finished. The chemical substances produced in the bodies of plants and animals remained just such "things-in-themselves" until organic chemistry began to produce them one after another, whereupon the "thing-in-itself" became a thing for us, as, for instance, alizarin, the colouring matter of the madder, which we no longer trouble to grow in the madder roots in the field, but produce much more cheaply simply from coal tar.[74]

In other words, practice is a kind of transformation of "things-in-themselves" into "things-for-us."

Lenin offers his own account of the relationship of practice and ideology in *Materialism and Empirio-Criticism*. In this work, Lenin's aim is to show that a recent trend in Russian Marxism, empirio-criticism (and its deviations, empirio-monism, empirio-symbolism, etc.), attempts to resolve the idealism/materialism dilemma through "experience," as per Ernst Mach, in light of certain scientific crises.[75] Practice for Lenin, much like Engels, involves the transformation of things in-themselves into things for-us and that practice is more fundamental than theory.[76] Again like Engels, Lenin maintains that philosophy ultimately falls into three different categories: materialism, idealism, and agnosticism.[77] Philosophy, insofar as it excludes or tries to usurp practice with theory, treats the absolute standpoint on material reality as one without the mark of human hands. Inevitably, this kind of materialism slips into agnosticism or idealism; that is, respectively, a skepticism toward knowledge of the material world or the annihilation of the material world altogether. For Lenin, both leave open the possibility for fideism, theism, and so on, that is, ideas that fundamentally preserve the existing social order.[78] He maintains that this process has repeated itself time and again throughout the history of philosophy, with the passage from materialism, to idealism, to agnosticism and back again and so on.[79] Lenin maintains that materialism is the superior term, insofar as the objectivity of the material world is guaranteed by the existence of the fossil record.[80] He maintains that the materialism of the philosophers, however, cannot properly avoid slipping back into idealism and agnosticism, so long as it ignores dialectics, according to which the relative standpoint and the absolute standpoint are one and the same. Dialectical materialism acknowledges the primacy of practice over theory and treats the relative, approximate character of all human knowledge by safeguarding itself from idealism and agnosticism and by effectively resolving the conditions that gave rise to the crisis of science.[81]

Perhaps the most significant of the aforementioned texts for Hessen as a Deborinite would have been Lenin's *Philosophical Notebooks*. Released for the first time in 1929, the notebooks are a collection of sketches on various philosophical works, the most significant of which are his "Conspectus on Hegel's Book *The Science of Logic*," "Conspectus on Hegel's Book *Lectures on the History of Philosophy*," and "On the Question of Dialectics," all three of which Lenin composed in Switzerland while in exile. Admittedly, Lenin does not always advance consistent positions throughout, but while he does make many similar claims in both the *Philosophical Notebooks* and in *Materialism and Empirio-Criticism*, he also appears to go beyond the limits of his previous works in crucial respects. The most important of these breaks, for our purposes, are his understanding of the notion of the thing in-itself/thing for-us and, consequently, on his notion of practice. In his defense of dialectical materialism, Lenin continues to stress the primacy of the material world with respect to cognition. However, one gets the impression that Lenin no longer understands "the world" in the same way as he once did, particularly when reading several remarks made throughout the course of the notebooks. He bases this claim on his understanding that for Hegel, "[t]he proposition that we do not know what Things-in-themselves are seems sagacious. The Thing-in-itself is an abstraction from all determination [Sein-für-Anderes {being-for-other}] [from all relation to Other], i.e., a Nothing. Consequently, the Thing-in-itself is 'nothing but an abstraction, void of truth and content.'"[82] In other words, while there cannot be a thing for-us without a thing in-itself, he now maintains that there cannot be a thing in-itself without a thing for-us either, given that a thing in-itself without appearance lacks determination, thereby making it effectively nothing. Instead, "[i]n life, in movement, each thing and everything is usually both 'in itself' and 'for others' in relation to an Other, being transformed from one state to the other."[83] Thus, "[t]he World in and for itself is identical with the World of Appearances, but at the same time it is opposite to it. . . . What is positive in the one is negative in the other."[84] The world in-itself is a moment of the world for-us, and vice versa and while the two are not identical, they do nevertheless form an inseparable unity in opposition. He identifies subjective and objective moments in the life of the human being that accompany this more nuanced distinction between the thing in-itself and the thing for-us. Accompanying this discovery is Lenin's development of his understanding of the relation between subject and object in practice. As in *Materialism and Empirio-Criticism*, Lenin defends the notion that "***[p]ractice is higher than (theoretical) knowledge,*** for it has not only the dignity of universality, but also of immediate actuality."[85] However, he no longer considers the human being simply as a subject and the natural world as the object. The objective world refers not only to the laws of the natural world, but also to purposive human activity. As he writes, "Two forms of the *o b j e c t i v e* process: nature (mechanical and chemical) and the *p u r p o s i v e* activity of man. The mutual relation of these forms."[86] This effort on behalf of human beings to self-actualize is an otherness within humanity, which strives to actualize itself by imposing itself upon the natural world. He goes on to say that "[i]n actual fact, men's ends are engendered by the objective world and presuppose it,—they find it as something given, present. But it *seems* to man as if his ends are taken from outside the world, and are independent of the world ('freedom')."[87] Lenin pinpoints a valuable tension here; on one hand, purposive, human activity is the effort to self-actualize, but on the other hand, human activity discovers its self-actualizations through the laws of nature. In light of this discovery, Lenin emphasizes the importance of studying human technique. As he writes, "[c]ontinuation of the work of Hegel and Marx must consist in the *d i a l e c t i c a l* elaboration of the history of human thought, science and technique."[88] He further states that "**IN HIS TOOLS MAN POSSESSES POWER OVER EXTERNAL NATURE, ALTHOUGH AS REGARDS HIS ENDS, HE FREQUENTLY IS SUBJECTED TO IT.**'"[89] In other words, the history of thought, science, technique, and tools bears witness to the history of the boundary between human beings and nature; this is the point at which purposive, human activity is articulated through the laws of nature and vice versa.

Elements from Marx, Engels, and Lenin are all present in Hessen's account, such that for Hessen, practice must also have had a twofold meaning. On one hand, the pursuit of science is the pursuit of understanding ever more complex forms of motion for Hessen. In this respect, he maintains that like Marx, Engels, and the Lenin of *Materialism and Empirio-Criticism*, practice is the transformation of the thing in-itself into a thing for-us. Hessen must maintain that as the arbiter of truth or falsity, practice shows that knowledge does not consist of the meditative gaze of an object but of its transformation. There is no such thing as knowledge of static objects because the instant that one seeks to know an object, one has already transformed it.

On the other hand, the pursuit of science for Hessen is the pursuit of self-consciousness. In this respect, and like the Lenin of the *Philosophical Notebooks*, practice is the transformation of a thing in-itself into a thing for-itself. Hessen must maintain that practice is the refinement of human purposive activity through the laws of nature, such that humanity becomes aware of itself through its own impositions upon nature. Thus, while the means of production and social relations constitute different moments of practice, each kind of practice contains its own division into two possible moments. The means and relations of production are both examples of practice, insofar as they transform things in-themselves into things for-us, even as they facilitate the transformation of things in-themselves into things for-themselves.

BUKHARIN'S "PRAXEOLOGY"

One last issue remains, however, concerning our understanding not only practice but also of ideology and the relationship of this pair for Hessen. In this concluding section of this chapter, then, I aim to concretize Hessen's understanding of practice and ideology with the controversial initiative of reading these notions through a work by his colleague, Bukharin. As indicated previously, once a high-ranking party official, Bukharin was also sent to the London conference in 1931, where his "Theory and Practice" was the first of many talks presented by the Soviet contingent and, in a way, provided the methodological groundwork for the papers that followed.[90] One could argue that assimilating Bukharin's and Hessen's views is fallacious because the former tended to associate more with the mechanists, whereas the latter sided with the Deborinites; put otherwise, the two men came from competing factions of dialectical materialist philosophers. However, Bukharin maintains that the objectivity of science cannot be understood as a disinterested pursuit of the truth, like Hessen. Moreover, although Bukharin was not a Deborinite, after Lenin's criticisms of his work for being too mechanistic, he undertook a close reading of Hegel and became an adept practitioner of dialectics.[91] With these overlaps, it is at least possible to state that one may be able to use Bukharin's account of the relationship between practice and theory to yield a better understanding of the relationship between practice and ideology in Hessen's 1931 Newton paper.

Bukharin begins "Theory and Practice" by elaborating, what he calls, "praxeology," or the study of the relationship between practice and theory, with the assumption that the former is more fundamental than the latter.[92] Bukharin defines practice as "an active break-through into reality, egress beyond the limits of the subject, penetration into the object, the 'humanising' of nature, its alteration,"[93] and explains that "[p]ractice is the refutation of agnosticism, the process of transforming 'things in themselves' into 'things for us,' the best proof of the adequacy of thought, and of its truth—understood historically, as a process."[94] This definition is accompanied by the observation that theory is simply "accumulated and condensed practice."[95] At first glance, Bukharin's account appears rather pedestrian: "[a]ction passes into cognition. Cognition passes into action. Practice drives forward cognition. Cognition fertilises practice. Both theory and practice are *steps* in the joint process of '*the reproduction of social life*.'"[96] He appears to be describing nothing more than the Baconian scientific method here: observation, induction, and deduction, and so on. It was not long until Bukharin began to tell a more complicated story, when he wrote

> [t]o the extent that it generalises the practice of material labour, and is qualitatively a particular and specific continuation of material labour, it is itself qualitatively a special, theoretical practice, to the extent that it is active (cf. e.g., the experiment)—practice fashioned by thought. On the other hand, practical activity utilises theory, and to this extent practice itself is theoretical.[97]

In other words, practice and theory do not form distinctively separable moments of a single process but, rather, practice consists of moments of theory, just as theory consists of moments of practice. The two are neither distinct, isolable processes, nor are they identical; rather, they are distinct aspects of a dialectical unity, and like not only Marx, Engels, and Lenin, but also like Hessen, Bukharin maintains that "their unity, develops on the basis of the *primacy of practice*."[98]

He defends the primacy of practice by arguing that "[t]he idea of the self-sufficient character of science ('science for science's sake')," or the belief that natural philosophy or science is a disinterested pursuit of the truth, independent of practice, "confuses the *subjective passions* of the professional scientists . . . with the objective *social rôle* of this kind of activity, as an activity of vast *practical* importance."[99] Science is never the simple pursuit of any one individual thinker, insofar as

> [a]ny empirical subject *always* goes beyond the bounds of "pure" sensual "raw material"; his experience, representing the result of the influence of the external world on the knowing subject in the process of his practice, stands on the shoulders of the experience of other people. In his "I" there is *always* contained "we." In the pores of his sensations there already sit the products of *transmitted* knowledge (the external expression of this are speech, language and conceptions adequate to words). In his *individual* experience there are included beforehand society, external nature and history—i.e., social history. Consequently,

> epistemological Robinson Crusoes are just as much out of place as Robinson Crusoes were in the "atomistic" social science of the eighteenth century.[100]

This dismantling of the private pursuit of knowledge illustrates the dependence of theory on practice in three respects:

> (1) *Historically*: the sciences "grow" out of practice, the "production of ideas" differentiates out of the "production of things"; (2) *sociologically*: "social being determines social consciousness," the practice of material labour is the constant "force mortice" of the whole of social development; (3) *epistemologically*: the practice of influence on the outside world is the primary "given quality."[101]

Each and every one of these remarks is in complete agreement with Hessen's criticism of ideological history that can be found at the beginning of the 1931 Newton paper.

Going beyond Hessen, however, Bukharin argues that the confusion between the "subjective passions of the professional scientists" and the "objective social role of this kind of activity" is the result of "a perverted ideological reflex of a society in which the division of labour has destroyed the visible connection between social function[s] (sic), separating them out in the consciousness of their agents as absolute and sovereign values."[102] In other words, the belief in the primacy of theory does not mean that theory has somehow usurped practice, nor does it mean that practice and theory have become divorced as, for Bukharin, this is simply impossible. Under the auspices of capitalism, economic production is characterized by both a division of mental labor from physical labor, as well as a radical division within different kinds of labor, thereby fragmenting pursuits of objective knowledge. From this isolation, each practice becomes an end in itself; the practice of theory is no exception.[103]

Returning to the 1931 Newton paper, the Hessen theses explain how socioeconomic roots, as means of production (machine technology) and relations of production (class standpoint), are on one hand, the foundation of Newton's natural philosophy, but on the other hand, they are the foundation only insofar as they are concealed as foundational. In the period of manufacture, economic production becomes disjointed from the perspective of society as a whole. Borrowing Bukharin's language, there was, as of yet, no subject that could account for socioeconomic roots as the conditions for objective science.[104] Newton's natural philosophy, then, is an example of mental labor and takes the pursuit of knowledge as an end in itself. Since it lacks the conditions for objective science, it can only compensate for this absence with the fetishization of machine technology and the class standpoint—in other words, by treating machine technology and the class standpoint as constitutive of knowledge itself. Thus, theory arises not only from practice but also from practice as concealed. One may call the specific difference between practice and theory in this context, ideology.

NOTES

1. This chapter is a revised version of my "Practice and Ideology in Boris Hessen's 'The Social and Economic Roots of Newton's *Principia*,'" *Telos* 190 (2020): 29–51, and also consists of passages from my "The Materialist Dialectic in Boris Hessen's Newton Papers (1927 and 1931)," *Historical Materialism* 28.4 (2020): 202–34 along with my "A Pantheon of Great Ideas: Boris Hessen and the History and Philosophy of Science."in *Manuscripts and Documents on the History of Physics: A Historical Materialist Textbook*, ed. Pietro Omodeo and Sean Winkler (Venice: Verum Factum, 2022), 35–73.

2. G. Freudenthal, "The Hessen-Grossmann Thesis: An Attempt at Rehabilitation," *Perspectives on Science* 15.2 (2005): 167; G. Freudenthal and P. McLaughlin, "Classical Marxist Historiography of Science: The Hessen-Grossmann-Thesis," in *The Social and Economic Roots of the Scientific Revolution: Texts by Boris Hessen and Henryk Grossmann*, ed. G. Freudenthal and P. McLaughlin (Dordrecht: Springer, 2009), 2–3.

3. Graham, "The Socio-Political Roots of Boris Hessen," 705–22; Schäfer, "Boris Hessen and the Politics of the Sociology of Science," 103–16.

4. C. D. Skordoulis, "Bukharin and the Social Study of Science," *Studies in East European Thought* 67 (2015): 76.

5. H. F. Cohen, *The Scientific Revolution: A Historiographical Inquiry* (Chicago: University of Chicago Press, 1994), 332. See also Freudenthal, "The Hessen-Grossmann Thesis," 167; Freudenthal and McLaughlin, "Classical Marxist Historiography of Science," 32.

6. J. P. Shapiro, "Rehabilitation Policy under the Post-Khrushchev Leadership," *Soviet Studies* 20 (1969): 490–98.

7. H. Grossmann, "The Social Foundations of the Mechanistic Philosophy and Manufacture," in *The Social and Economic Roots of the Scientific Revolution: Texts by Boris Hessen and Henryk Grossmann*, ed. G. Freudenthal and P. McLaughlin and trans. G. Shalit (Dordrecht: Springer, 2011), 103–56; H. Grossmann, "Descartes and the Social Origins of the Mechanistic Concept of the World," in *The Social and Economic Roots of the Scientific Revolution: Texts by Boris Hessen and Henryk Grossmann*, ed. G. Freudenthal and P. McLaughlin and trans. G. Shalit (Dordrecht: Springer, 2011), 157–230; H. Grossmann, "Additional Texts on Mechanism," in *The Social and Economic Roots of the Scientific Revolution: Texts by Boris Hessen and Henryk Grossmann*, ed. G. Freudenthal and P. McLaughlin and trans. G. Shalit (Dordrecht: Springer, 2011), 231–38.

8. Freudenthal and McLaughlin, "Classical Marxist Historiography of Science," 1.

9. Ibid.

10. J. M. Bochenski, *Soviet Russian Dialectical Materialism [Diamat]* (Dordrecht: D. Reidel Publishing Company, 1963), 34.

11. Ibid.

12. V. I. Lenin, "On the Significance of Militant Materialism," in *V.I. Lenin – Collected Works, Vol. 33*, ed. D. Skvirsky and trans. G. Hanna (Moscow: Progress Publishers, 1973 [1922]), 233–34. See also Josephson, *Physics and Politics in Revolutionary Russia*, 246; Schäfer, "Boris Hessen and the Politics of the Sociology of Science," 114.

13. Graham, "The Socio-Political Roots of Boris Hessen," 709; N. Krementsov, *Stalinist Science* (Princeton: Princeton University Press, 1997), 31–32.

14. Bochenski, *Soviet Russian Dialectical Materialism*, 35–36.

15. М.Б. Митин, В.Н. Ральцевич and П.Ф. Юдин, «О новых задачах марксистско-ленинской философии», Правда, 7 June 1930; Б.М. Гессен, et al. «О борьбе на два фронта в философии», Под знаменем марксизма 5 (1930): 139–49.

16. Гессен, et al. «О борьбе на два фронта в философии», 141–42.

17. J. Stalin, *Dialectical and Historical Materialism*, trans. Unknown (New York: Prism Key Press, 2013 [1938]), 7.

18. Гессен, et al. «О борьбе на два фронта в философии», 139, 141. See also Б.М. Гессен and И.И. Подволоцкий, «Философские корни правого оппортунизма», Под знаменем марксизма 9 (1929): 1–3.

19. A. Pope, "Epitaph on Sir Isaac Newton," in Every Day in the Year: A Poetical Epitome of the World's History, ed. J. L. Ford and M. K. Ford (New York: Dodd, Mead and Co., 1902 [1730]), https://www.bartleby.com/297/154.html. See also Hessen, "The Social and Economic Roots of Newton's *Principia*," 41.

20. A.N. Whitehead, "The First Physical Synthesis," in *Science and Civilization*, ed. F. S. Marvin (London: Oxford University Press, 1923), 162. See also Hessen, "The Social and Economic Roots of Newton's *Principia*," 41. Freudenthal and McLaughlin indicate that this is, in fact, a paraphrasing of Whitehead's actual remark which reads as follows: "our modern civilization is due to the fact that in the year when Galileo died, Newton was born. Think for a moment of the possible course of history supposing that the life's work of these two men were absent." Hessen, "The Social and Economic Roots of Newton's *Principia*," 96n3.

21. Hessen, "The Social and Economic Roots of Newton's *Principia*," 61.

22. Ibid., 44.

23. Ibid., 42.

24. Ibid. See also V. I. Lenin, *Karl Marx: A Brief Biographical Sketch with an Exposition of Marxism*, in *V.I. Lenin: Collected Works, vol. 21*, ed. S. Smith and trans. C. Dutt (Moscow: Progress Publishers, 1974 [1914]), 56.

25. Hessen, "The Social and Economic Roots of Newton's *Principia*," 41–42.

26. Ibid., 43.

27. Ibid.

28. Ibid., 43–44, 45.

29. Ibid., 76. See also K. Marx, *Capital, Vol. 1*, in *Marx and Engels – Collected Works, Vol. 35*, ed. A. Chepurenko and trans. Y. Vashchenko and L. Zalunina (London: Lawrence and Wishart, 2010 [1867]), 343.

30. Hessen, "The Social and Economic Roots of Newton's *Principia*," 44. See also Marx, *Capital, vol. 1*, 356.

31. Hessen, "The Social and Economic Roots of Newton's *Principia*," 44. See also Marx, *Capital, vol. 1*, 344.

32. Hessen, "The Social and Economic Roots of Newton's *Principia*," 45.

33. Ibid., 45–52.

34. Ibid.

35. Ibid., 53.

36. Ibid., 52.

37. Ibid., 53–57.

38. Ibid., 68–69.

39. Ibid., 70.

40. Ibid., 66.

41. Ibid., 65.

42. Ibid., 61.

43. Ibid., 62.

44. Ibid., 72.

45. Ibid., 66.

46. Ibid.

47. Ibid.

48. Ibid., 64.

49. Ibid., 67.

50. Ibid., 69.

51. Ibid., 74.

52. Ibid., 75–75, 79–80.

53. Ibid., 82–89.

54. Ibid., 88.

55. Ibid., 88–89.

56. Ibid., 44, 59, 96n9.

57. Ibid., 49, 96n15.

58. Ibid., 56, 97n26.

59. Ibid., 55, 97n25.

60. Ibid., 65, 70, 98n40, 98n41, 99n61.

61. Ibid., 63, 97–98n37.

62. Ibid., 62, 66, 67, 97n35, 97n36, 97–98n37, 99n56, 99n57.

63. Ibid., 61, 66, 97n34, 99n51.

64. Ibid., 76, 77, 78, 100n69, 100n73, 100n74.

65. Ibid., 75, 100n68.

66. Ibid., 82–89.

67. Ibid., 81, 101n81.

68. K. Marx and F. Engels, *The German Ideology*, in *Marx and Engels – Collected Works, Vol. 5*, ed. J. Cohen, et al. and trans. W. Lough (London: Lawrence and Wishart, 2010 [1845]), 59; K. Marx, *A Contribution to the Critique of Political Economy*, in *Marx and Engels – Collected Works, Vol. 29*, ed. J. Cohen, et al. and trans. V. Schmittke and Y. Sdobnikov (London: Lawrence and Wishart, 2010 [1859]),

263. See also Hessen, "The Social and Economic Roots of Newton's *Principia*," 43, 87, 96n6, 101n86.

69. Hessen, "The Social and Economic Roots of Newton's *Principia*," 81.

70. Ibid. See chapter 4.

71. Ibid., 41, 71, 74, 82, 88.

72. K. Marx, "Theses on Feuerbach," in *Marx and Engels – Collected Works vol. 5*, ed. J. Cohen, et al., and trans. W. Lough (London: Lawrence and Wishart, 2010 [1845]), 5.

73. Ibid., 3.

74. F. Engels, "Ludwig Feuerbach and the End of Classical German Philosophy," in *Marx and Engels, Collected Works, vol. 26*, ed. B. Tartakovsky and trans. N. Jacobs, et al. (London: Lawrence and Wishart, 2010 [1886]), 367–68.

75. V. I. Lenin, *Materialism and Empirio-Criticism: Critical Notes concerning a Reactionary Philosophy*, in *Collected Works of V.I. Lenin, Vol. 13*, ed. A. Trachtenberg and trans. D. Kvitko (New York: International Publishers Co., Inc., 1927 [1908]), 5.

76. Ibid., 77–78, 83, 109–14.

77. Ibid., 14–15.

78. Ibid., 149, 190.

79. Ibid., 23, 46–47, 102. See also L. Althusser, "Lenin and Philosophy," in *Lenin and Philosophy and Other Essays*, trans. B. Brewster (New York: Monthly Review Press, 1971 [1968]), 54, 64.

80. Lenin, *Materialism and Empirio-Criticism*, 49, 52, 96, 104, 154.

81. Ibid., 107–8, 221, 262.

82. V. I. Lenin, "Conspectus of Hegel's Book *The Science of Logic*," in *V.I. Lenin – Collected Works, Vol. 38*, ed. S. Smith and trans. C. Dutt (Moscow: Progress Publishers, 1972 [1915]), 108–9.

83. Ibid., 109.

84. Ibid., 152–53.

85. Ibid., 213.

86. Ibid., 187–88.

87. Ibid., 189.

88. Ibid., 147.

89. Ibid., 189.

90. Skordoulis, "Bukharin and the Social Study of Science," 76.

91. Ibid., 81.

92. N. Bukharin, "Theory and Practice from the Standpoint of Dialectical Materialism," in *Science at the Cross Roads – Papers Presented to the International Congress of the History of Science and Technology, Held in London from June 29th to July 3rd, 1931 by the Delegates of the USSR*, ed. N. Bukharin (London: Frank Cass and Co. Ltd., 1971 [1931]), 14, 16.

93. Ibid., 17.

94. Ibid.

95. Ibid., 13.

96. Ibid., 14.

97. Ibid., 13–14.
98. Ibid., 14–15.
99. Ibid., 20.
100. Ibid., 12. See also Marx, *Capital, vol. 1*, 87–88.
101. Bukharin, "Theory and Practice from the Standpoint of Dialectical Materialism," 15.
102. Ibid., 20.
103. Ibid., 30–33.
104. Skordoulis, "Bukharin and the Social Study of Science," 87.

Chapter Two

Idealism and Materialism[1]

This second chapter proceeds by discussing Boris Hessen's understanding of the role played by ideology in the development of the sciences, specifically through the dynamic between the philosophical persuasions of "idealism" and "materialism." It should be borne in mind that Hessen published an article titled "Preface to Articles by A. Einstein and J.J. Thomson" (hereafter, the 1927 Newton paper) in the Soviet journal, *Under the Banner of Marxism* four years prior to his presentation of the 1931 Newton paper; the former was a piece that he wrote in commemoration of the bicentennial of Sir Isaac Newton's death.[2] While both papers discuss Newton, the former bears a much stronger resemblance to works found in the so-called "internalist" camp of the history and philosophy of science, which was the very method that Hessen had criticized in 1931.[3] For decades, scholars have puzzled over why Hessen chose to argue for seemingly diametrically opposed positions. I respectfully maintain, however, scholars have yet to provide a satisfactory explanation as to why he would have done so.

In this chapter, I claim that Hessen does not actually advance two different methods but, rather, a single, dialectical materialist approach that accounts for the "unity in opposition" of the external and internal dimensions of natural scientific theory. While the role of practice and ideology in the formation of natural scientific theory are the underlying themes of the 1931 text, he devotes his attention to the interplay between the philosophical persuasions of materialism and idealism as part of the ideological struggle in the development of natural scientific theory in the 1927 work. For Hessen, understanding this dynamic is necessary to show that dialectical materialism is actually not meant to offer a final, monolithic natural scientific theory but is instead meant to offer a method for negotiating the complexity of how new scientific theories emerge. I develop this position throughout the course of three sections. In the first section, The Newton Papers, I provide general summaries of Hessen's two papers on Newton. In the second section, The Hessen Paradox, I present and criticize three prevalent hypotheses found in the secondary literature that attempt to resolve the discrepancy between Hessen's works. In the third section, Ideological Struggle, I argue that while Hessen discusses the same topics in both 1927 and in 1931, in different respects, he advances two distinct layers of the same dialectical materialist approach in both; on one hand, by showing the socioeconomic roots of the natural sciences, while on the other hand, also showing the way in which idealistic theories facilitate the development of new paradigms of materialism. To support my claims, I draw from Hessen's two papers about Newton, his contributions to *Under the Banner of Marxism* as well as some selections from the secondary literature.

THE NEWTON PAPERS

For this first section, I will establish the basic arguments outlined in Hessen's two papers that are devoted to Newton. As indicated, Hessen's 1931 Newton paper is often credited for being the inaugural and paradigmatic example of the externalist approach to the history and philosophy of science. The *Dictionary of the History of Science* defines externalism as "[t]he view that social, political, and economic circumstances affect the pursuit of knowledge of Nature."[4] In 1927, however, Hessen wrote the "Preface to Articles by A. Einstein and J.J. Thomson," which was an attempt to grapple with Einstein's, Thomson's, and Horace Lamb's assessments of Newton's legacy in light of the crisis of physics in the early twentieth century and the emergence of quantum mechanics. The paper is often treated as an example of internalist history and philosophy of science, which the *Dictionary of the History of Science* defines as "[t]he view that science is primarily an abstract intellectual enterprise insulated from social, political, and economic circumstances."[5] In what follows, I will provide summaries of Hessen's two works on Newton to establish a basis on which they can be compared and contrasted, an operation that will occur later on in the chapter.

The 1931 Paper

While Hessen's text from 1931 is the chronologically later work, I will summarize it first as it is the paper with which most scholars are most familiar. Although this paper has already been discussed in the previous chapter, I will briefly reacquaint the reader with its essential arguments for the sake of juxtaposing it with Hessen's other piece.[6] We might recall that Hessen begins the 1931 paper by arguing that Newton's *Principia* had traditionally been read as the fruit of Newton's profound genius. Although Newton's genius goes unquestioned, according to Hessen, this approach to the history and philosophy of science misconstrues the origin of the theoretical problems with which Newton grapples in the first place. Putting Newton in a larger context, he states, one cannot ignore that the natural sciences of the sixteenth and seventeenth centuries in general, and Newtonian physics in particular, are the result of changes found in the socioeconomic and technological conditions of the day.[7]

Hessen offers his four, supporting subtheses in order to defend the overall thesis. In subthesis 1, "The Economics, Technology and Physics of Newton's Era/The Physical Themes of the Era and the Contents of the *Principia*," Hessen proposes that during Newton's lifetime, there was an increased significance in the role played by machines in early capitalist manufacture, particularly in the spheres of communication, industry, and warfare. Unable to discuss matters pertaining to the development of machines in universities, the rising bourgeoisie/merchant class, who increasingly controlled the means of production, founded scientific societies within which they were able to discuss and keep track of them. This established a context from which practical problems were transmitted across different industries, giving rise to abstract, theoretical problems that eventually constituted the basis of classical mechanics. Newton's exposure to classical mechanics was no accident, given that he was a member of the most notable scientific society of the day and the first of its kind, namely, the Royal Society. For Hessen, the theoretical problems found within Newton's mechanics are directly traceable to practical problems that derive from the study and improvement of machines for early capitalist manufacture.[8] According to subthesis 2, "The Class Struggle during the English Revolution and Newton's Worldview," Newton's system is not wholly mechanistic but is characterized by a unique "division of labor": On one hand, Newton affords mechanistic causality a prominent role in nature, but on the other hand, he also upholds the concepts of so-called "action at a distance" and the providential notion of the divine. For Newton, these elements are organically bound together, but as Hessen shows, they need not necessarily be so. He explains that other mechanistic philosophers writing at the time, for instance, either advanced entirely different conceptions of the divine (i.e., René Descartes) or omitted the concept of the divine altogether (i.e., Thomas Hobbes, Robert Overton). Newton's particular conception is the product, Hessen states, of the influence of his bourgeois class standpoint, which sought democratization and religious tolerance during the so-called "class compromise" in late seventeenth-century England.[9] According to subthesis 3, "Engels' Conception of Energy and the Lack of the Law of Energy

Conservation in Newton," Hessen raises the question as to why Newton limited himself to understanding motion as mechanical, when there are, in fact, various different kinds of motion that can be observed in nature. He maintains that this limitation was not the product of some failure of Newton's genius but was rather a limitation of the kinds of motion exhibited in the contemporary means of production. Other kinds of motion would not be discovered until the rise of thermodynamics, which Hessen explains as having arisen from the incorporation of the steam engine into the factory setting and the need to both study and develop it. Newton could not possibly have theorized motion as anything other than mechanical and as the means of production through which the founding principle of thermodynamics, the law of energy conservation, had not yet arisen.[10] In subthesis 4, "The Machine-Wreckers in Newton's Age and the Present-Day Wreckers of the Productive Forces," Hessen argues that the tactic of machine-wrecking in the protest of job loss to mechanization is complicit with so-called "machine-fetishism," a central component of bourgeois ideology by which machines are seen as being in competition with human labor. He maintains that the proletariat must see past machinery as the threat to their livelihood and ought to identify its true source as the social relations themselves, that is, the class hegemony of the bourgeoisie and, broadly speaking, class society as such. Once they transform social relations, they will also transform their perspective on the means of production and, with both having been accomplished, establish an entirely new basis for future scientific endeavors.[11]

The 1927 Newton Paper

Hessen's 1927 article is a preface to Russian translations of Albert Einstein's "The Mechanics of Newton and Their Influence on the Development of Theoretical Physics,"[12] J. J. Thomson's "Newton's Work in Physics,"[13] and Horace Lamb's "Newton's Work in Mechanics"[14] for *Under the Banner of Marxism*; all of these texts were written in commemoration of the bicentennial of Newton's death.[15] In the preface, Hessen notes that the occasion was opportune for reflecting upon the renewed interest in Newton's work at the time; after all, this was a trend which he attributed to the ensuing crisis in early twentieth-century physics.[16] This crisis, Hessen claims, was the result of new scientific discoveries that had challenged the foundations of Newtonian physics and with it, many felt, the backbone of physics itself, that is, the law of causality. For Hessen, the crisis did not mean dispensing with this law altogether but, instead, involved looking beyond its conception in Newtonianism.[17] Just as it had previously, physics was undergoing a transformation; as the concept of causation in Scholastic Aristotelian physics had been overcome by that of Newtonian physics, so too was the latter now being overcome by its conception in quantum mechanics. Quantum mechanics could not, however, surmount the crisis of physics on its own, given that this would require a dialectical materialist reading that could "replac[e] the metaphysical opposition of necessity and chance with the dialectical concept of causality."[18] I will set aside Hessen's dialectical materialist reading of quantum mechanics until the next chapter to keep the focus on his usage of this method in formulating his subthesis on Newton, which supports the general position that "[t]he development of science . . . . is inextricably linked with the development and change of basic methodological concepts."[19]

For this subthesis, Hessen argues that Newtonian physics had always been an incomplete system of physical causality and that it had become untenable; the reason provided was that it had been extended beyond its proper domain. The notion that Newtonian physics is complete, he suggests, is by and large the product of the work of subsequent generations who sought to render Newton's thought into a system. In Newton's own writings, however, "we do not find a philosophically grounded and consistently conducted natural-scientific worldview,"[20] given that he expressly does not espouse a comprehensive, universal brand of causality, but instead advances two, distinct kinds. In a letter to Newton from June 29, 1686, cited by Hessen, Edmond Halley summarizes this distinction by indicating that "[t]he main idea of the '*Principia*' is to present the motion of the planets as a consequence of the combination of two forces: 'a composition of a Descent towards the sun, and an imprest motion,'"[21] that is, action at a distance/divine causation and mechanical motion, respectively. For Newton, these two conceptions of causation are "organically linked," but they are, in fact, quite separable.[22] Hessen characterizes the synthesis between the two as a "peculiar 'division of labour'"[23] within Newton's system, which is not the sole, possible rendering of mechanical causation but is an approach that was typical among English thinkers in Newton's day for rendering the natural sciences amenable to religious dogma.[24] He writes that "[o]n English soil, mechanical determinism is generally recognized, but it is often intertwined with religious dogma."[25] The incompleteness of Newton's worldview, Hessen explains, is due to the fact that it was a product of the history of the natural sciences at a certain phase of their development. He writes that "Newton's physics, like any methodology, is a necessary stage in the development of natural science; one whose subsequent development more and more reveals the insufficiency and unsatisfactory nature of its methodological premises."[26] To say that it marks a particular phase of development has the same effect as saying that Newtonian physics cannot be properly understood independently of its relationship with other moments in the history of the natural sciences, both diachronically and synchronically.

Diachronically, the new physics rose to prominence in "[t]he 17th century[,] . . . . beginning with Galileo, [and in] the struggle against Scholastic Aristotelian physics."[27] This conflict was waged in large part against, what was taken to be at the time, causal explanations via the *qualitas occulta* [hidden quality], which were increasingly seen as problematic because they were thought to have an entirely tautological nature; for instance, "a magnet attracts because it possesses the magnetic force to pull."[28] The opposition to this concept laid the groundwork for modern natural science.[29] Hessen also argues, drawing from the work of J. J. Thomson, that synchronically, Newtonian physics cannot be properly understood independent of its opposition with other emerging paradigms in the natural sciences of the time, most importantly, Cartesian physics.[30] He writes that "[a]ny transitional epoch must be considered in the whole variety of its contradictory trends. The figure of Luther is[, for instance,] incomplete and incomprehensible without the figure of Müntzer."[31] The struggle against the *qualitas occulta*, then, was not a unified effort, but was instead a struggle bifurcated from within, whose "main question . . . was [that] of the general method of scientific research."[32] Further on, he writes that "[t]he reaction against Scholastic physics is fueled by two sources: Bacon's empiricism and the mechanistic principles of Descartes' physics."[33] The characteristic difference between the two is that "[i]f Bacon chose experience as the primary starting point, then according to Descartes, the only true path to knowledge is deduction."[34] Newtonian physics arose from the context of the struggle between both of these two methodologies. While the title of Newton's magnum opus, *Principia*, echoes that of Descartes, it cannot be treated as a synthetic work that combines empiricist and rationalist methodology. Rather, "Newton's methodology is that of empiricism, which is wrapped in a mathematical form . . . .[; his] methodological reasoning does not constitute a whole system and does not serve as the starting point for his constructions,"[35] while "Descartes builds his physics into a complete, rationalistic system, sometimes even at odds with the known facts of the time."[36] Thus, even though Newton and Descartes both oppose Scholastic Aristotelian physics, they also oppose each other by adopting competing methodologies.

Competing conceptions of causation follow on from these opposing methodologies. In Newtonian physics, causal explanations are derived from experience, which are then interpreted according to a "mathematical phenomenology;" that is, phenomena are extrapolated from experience and rendered into mathematical notation.[37] Newton's primary source for the observation of phenomena was the continuous motion of the planetary orbits. He translated these observations into universal mathematical formulas, thereby applying the laws of the macrocosm to the microcosm.[38] From this method, Newton concludes that there must be two kinds of causation at work in the universe:

mechanical motion and action at a distance/divine causation. In Cartesian physics, however, causal explanations are derived from deduction through the rationalistic method. Descartes follows a priori concepts to the conclusion that causation in nature can *only* be mechanical. In support of this position, he writes the following in his *Principles of Philosophy*:

> I recognize no matter in corporeal things apart from that which the geometers call quantity, and take as the object of their demonstrations, i.e. that to which every kind of division, shape and motion is applicable. Moreover, my consideration of such matter involves *absolutely nothing* apart from these divisions, shapes and motions; and even with regard to these, I will admit as true only what has been deduced from indubitable common notions so evidently that it is fit to be considered as a mathematical demonstration. And since all natural phenomena can be explained in this way, as will become clear in what follows, I do not think that any other principles are either admissible or desirable in physics.[39]

Descartes maintains that the laws of the macrocosm and the microcosm may not be the same by strictly abiding by the rationalistic method and by a single form of causation, while the macrocosm and the microcosm both abide by the same laws for Newton. The consequence was, Hessen writes, that because "Newton's physics is molar (macroscopic) physics *par excellence*[, i]t is absolutely clear that with the transition to the study of the microcosm, a new approach to the study of phenomena is required."[40] The whole Newtonian paradigm was called into question because Newton and his followers had assumed that causation entailed continuity and even the unity of the laws of the macrocosm and the microcosm, with the observation of discontinuous process at the microcosmic level late in the nineteenth and early in the twentieth centuries.

THE HESSEN PARADOX

This chapter's second section proceeds by explaining what I refer to as the Hessen paradox, or the apparent tension found in Hessen's methodology concerning the history and philosophy of science. The two papers outlined previously appear to highlight a fundamental conflict that comprises at the heart of Hessen's thought: On one hand, we find one of the (if not *the*) founding document of the externalist approach to the history and philosophy of science, but on the other hand, we also find a seemingly conventional, internalist approach to that field. The secondary literature has yielded three hypotheses that attempt to account for this paradox: first, the Graham thesis; second, the dialectical ontological thesis; and third, the Machist thesis. This section will describe each of these hypotheses, followed by a critique and a distillation of their common insights.

The Graham Thesis

In this section, I will explore what I will refer to as the Graham thesis, as presented by historian of science, Loren Graham in an article from 1985, titled "The Socio-Political Roots of Boris Hessen." In this work, Graham argues that Hessen does not in fact adopt different points of view of the two Newton papers. Instead, he maintains that in both texts, "Hessen wished to differentiate between the social origins of science and its cognitive value;"[41] that is, Hessen espouses an internalist approach through and through—in an explicit manner in most of his works and in an implicit manner in 1931—remaining committed to the defense of the natural sciences from "ideological perversion."[42] Graham claims that Hessen's decision to adopt allegedly vulgar Marxist rhetoric in the 1931 paper would have been to appease the Stalinist camp of dialectical materialists.

Graham begins his analysis by contending that when reading Hessen's externalist text, we should not exempt it from being subject to an externalist analysis itself; that is, just as Hessen took Newton's context into consideration, so too must Hessen's context be examined alongside his ideas.[43] For Graham, one cannot properly understand Hessen's 1931 text independently of his flourishing during the late 1920s and the NEP, as well as the drastic turn in the Soviet Union's policies after the Great Break.[44] Where Hessen had previously been freer to articulate his own position, the political climate would have become increasingly prohibitive by the time he wrote his 1931 paper, meaning that he would have been forced to adopt the language of Stalinist Marxism to show his adherence to the Communist Party.

However, as Graham notes, behind the rather dogmatic language of the 1931 work, Hessen offered a carefully constructed critique of the ideological corruption of the natural sciences which had been occurring in the Soviet Union; this is a position that, he maintains, is fully in line with that of his earlier writings. Graham argues that in the 1931 Newton text, Hessen covertly defended an internalist conception of the development of the natural sciences, one based upon the following analyses. First, he maintains that Hessen's 1931 paper differs radically not only from his 1927 text, but also from every other work found in his oeuvre. As he states,

> [t]here [in the Soviet Union] he did not produce articles about the social context of science, but instead was engaged in a spirited defence of relativity theory and quantum mechanics against vulgar Marxist critiques of these revolutionary developments in physics. In these articles he maintained that a separation could

> be made between the intellectual content of a theory and the social context in which it was produced, a view that sounds very similar to the one which the more outspoken external historians of science in the West, citing his work, would eventually question.[45]

He elaborates upon this claim by stating that

> [w]hen reading the [1931] essay I have always been struck by its flamboyant application of an elementary form of Marxism to social and intellectual relations. Especially strong is the emphasis on technology as a formative force in intellectual life. After studying Hessen's other articles written in the Soviet Union, at first I had difficulty recognizing the same author. Normally, he was a sober and careful scholar writing on such topics as probability theory or relativistic mechanics. In other articles he stressed logical and philosophical themes, not technology.[46]

Graham corroborates his take with that of Ernst Kolman, the Communist Party Secretary and a once loyal Stalinist, who was assigned to oversee the Soviet delegation to the London conference in 1931.[47] Graham notes that "Kolman agreed . . . that Hessen's London paper was atypical of his interests."[48]

Second, according to Graham, Hessen was subject to considerable scrutiny by the Communist Party prior to his attendance at the London conference; this is a fact of which Hessen is believed to have been well aware.[49] Graham maintains that Hessen came under the increasingly watchful eye of the state after the Great Break because he was a member of the Deborinite school of dialectical materialists.[50] He is thought to have attracted even more suspicion than his fellow Deborinites because he had been an ardent proponent of quantum mechanics and relativity theory in previous years.[51] At the time, these two developments in the natural sciences were seen as "bourgeois," "Menshevist idealistic," "Machist," and "Right deviationist" by Stalin and his sycophants.[52] It is quite likely, Graham argues, that when Hessen was chosen to be a member of the Soviet delegation to the London conference in 1931, he was in fact being subjected to a test of his loyalty.[53] Kolman substantiates this position by noting that the Communist Party members of the delegation "had been instructed by no less an important a body than the Politburo to emphasize Marxism in their reports,"[54] and were, more specifically, told to heed Stalin's declaration that "technology in the current stage decides everything."[55] Graham argues, then, that the change in Hessen's position must be understood in relation to his effort to placate party authorities—hence, the influence of external, "socio-political" factors.[56]

Third, Graham argues that while Hessen adopted language that would, at least temporarily, appease Stalin and his followers, the externalist position that he advances is actually deliberately self-undermining upon closer scrutiny. As indicated previously, for Graham, the common theme that runs throughout Hessen's entire oeuvre is his defense of the natural sciences from ideological perversion. "Deep down," Graham writes "there is a continuous theme in Hessen's London paper and his Soviet articles: the unlinking of the value of a scientific theory from the ideological framework in which it developed."[57] He elaborates this point in the following way:

> [t]he overwhelming impression I gain from the London paper is that Hessen had decided "to do a Marxist job" on Newton in terms of relating physics to economic trends, while imbedding in the paper a separate, more subtle message about the relationship of science to ideology. He must have realized that by interpreting Newton in elementary Marxist economic terms, he could accomplish two important goals: first of all, he could demonstrate his Marxist orthodoxy, something being seriously questioned by his radical critics back in the Soviet Union; second, he could, by implication, defend science against ideological perversion by pointing to the need to separate the great merit of Newton's accomplishments in physics from both the economic order in which they arose and the philosophical and religious conclusions which Newton and many other people drew from them. . . . Hessen was illustrating that Marxists should simultaneously recognize the value of Newton's physics while seeing that it developed in mercantilist England and was used as a tool to support religion; therefore, they should similarly recognize the value of Einstein's and Bohr's physics while acknowledging that they arose in imperialist Europe and are often used to counter Marxism. . . . His performance carried the implicit message to Bolshevik critics of relativity physics: "What you do to Einstein and Bohr, I can do to Newton; so let's leave the physics alone."[58]

In other words, Hessen meant to capitalize upon the popularity of Newtonian physics in the Soviet Union at the time to show that it was a physical system that was ultimately divisible into two parts: One part marked the advancement of classical mechanics and another part stood in the way of that same progress, the former being internal to and the latter being external to the sciences. Graham concludes, then, that "[t]he unwritten final line [of the 1931 paper] was that when Einstein wrote on religion or philosophy he also merely expressed his social context and therefore these views should not be held against his physics."[59] Hessen's externalist standpoint is, for Graham, simply a foil for illustrating how social factors actually corrupt scientific progress.

Graham is not alone in defending this position, given that it can also be found in a lesser-developed form in the works of David Joravsky and Wolf Schäfer. Joravsky maintains that

> [a]s possible reasons for the drastic difference between the two essays one may suggest: first, Hessen's basic purpose, before the English audience, of using

> Newton merely as a peg on which to hang a lecture on basic Marxist social theory; second, the recent condemnation of the Deborinite School, of which Hessen was a leading member.[60]

Schäfer similarly comments that "there seems to be enough evidence that Hessen's London thesis was rooted in an externally induced political learning process, to put it mildly."[61] It is for these reasons, according to the Graham thesis, Hessen's 1931 paper deconstructs the externalist approach to the history and philosophy of science. His position remained fundamentally internalist throughout the entirety of his career because Hessen's aim in 1931 had been to criticize the ideological corruption of the sciences.[62]

The Dialectical Ontological Thesis

I will now proceed by discussing the so-called "dialectical ontological" thesis, as presented by historian and philosopher of science Olga Stolyarova in her article titled "Boris Hessen's Ideas and Russian Philosophy" from 2017. In this work, Stolyarova argues that while Hessen does appear to favor an internalist approach throughout his oeuvre and an externalist approach in 1931, he would have never seen these positions as being mutually exclusive. Ultimately, she argues, Hessen espouses a view that she describes as being dialectical ontological, or that according to which there exists a "complex relationship between the historical *a priori* social, ideological and technical, on the one hand, and the cognitive content of scientific theories."[63] In other words, she maintains that Hessen understands the history of science as unfolding from a complex interplay of socioeconomic, technological, and ideological factors, all while affording an equal status to material and ideal causes.

Stolyarova begins her argument by pointing out that in both papers Hessen treats materialism not as a static philosophical persuasion but, rather, as something that is subject to perpetual revision over the course of history.[64] She writes that for Hessen, "the dialectical understanding of science and its history makes it possible to take into account concrete historical forms of materialism, without dogmatizing any of them."[65] By characterizing materialism as subject to perpetual revision, Hessen effectively undermines the distinction between materialism and idealism altogether. With a transient character, the very features that identify a certain position as materialistic at one moment may identify that same position as idealistic in another, and vice versa. Consequently, Hessen does not favor materialism or idealism at all, but the term that unites them, namely, dialectics. As she writes, "the terms 'materialism' and 'idealism', separating philosophers on opposite sides of the barricades, are not as important as the concept of 'dialectical ontology' which unites them."[66] Hessen's overall position, then, is better defined as such: "[t]he meaning of Hessen's concept, as well as the meaning of the concepts of those philosophers who continue and develop the tradition of dialectical ontology, consists in the fact that by introducing the historical *a priori* of social things, the context of discovery is brought into 'dynamic balance' with the context of justification."[67] Moreover, she states that Hessen does not misunderstand his own position, but instead adopts a position with less of an affinity toward Marxism and more of an affinity for the history of Russian philosophy. She states that "[t]he Russian philosophical tradition, in which dialectical ontology plays an important role, is the closest context of the Hessen report and the ideas expressed in it."[68] Hessen is neither an internalist or externalist, nor an idealist or materialist, but a dialectician, unlike Marx, Engels, and Lenin and more like Pavel Florenskii, Evald Ilyenkov, and Aleksei Losev.[69]

The Machism Thesis

Finally, I will present what I refer to as the Machism thesis, as presented by historians and philosophers of science, Gerardo Ienna and Giulia Rispoli in their 2019 article, titled "Boris Hessen at the Crossroads of Science and Ideology." In this piece, Ienna and Rispoli argue that despite the fact that Hessen often claimed to be a proponent of dialectical materialism, the positions that he defends in his work were neither dialectical nor materialist.[70] Rather, they maintain that Hessen was an adherent of Machist epistemology through and through. As they write, "Hessen's views were closer to an understanding of materialism and scientific development that were rooted in Russian Machism rather than in dialectical materialism."[71] They defend their argument on the bases, first and foremost, that his positions were often more consistent with Machism than dialectical materialism and, second, that he was highly influenced by Richard von Mises, who was himself an adherent of Machist epistemology.

They provide a whole host of reasons as to why many of the positions that Hessen defended are more consistent with Machism than with dialectical materialism. For instance, they explain that Hessen did not uphold the concept of the so-called unity of opposites—the central tenet of dialectics—for studying phenomena, but rather that of the "theory of equilibrium." They contend that "[h]is [Hessen's] ideas about the relationship between structure and superstructure are more akin to Bogdanov's theory of equilibrium which in turn inspired Nikolai Bukharin's account of the complementary relationship between theory and praxis."[72] Moreover, they assert that Hessen was ultimately indifferent to the distinction between the philosophical persuasions of materialism and idealism, given that his primary focus lay on the concept of historical development. They note, for instance, that "[t]he importance of history for the understanding of scientific development as an open process in which material and ideal elements are interwoven dissociates Hessen's view from dialectical materialism."[73] They contend that "Hessen seems to defend an *empiriomonistic* view of knowledge production that saw cognition as a historical process in Bogdanovite terms."[74] Like Bogdanov, Hessen believed that the historical process, by which theories develop, is no less important than the theory in its present form. For instance, in his *Cognition from a Historical Point of View*, Bogdanov writes that

> [t]he history of the development of the human mind is an uninterrupted process of metamorphoses and transformations of old ideas into new ones, and there is no doubt that at every stage of understanding any new "truth" sooner or later will turn wrong. But this does not characterize only history. Actually, even current ideas represent a struggle between "old truths" and the "new ones." . . . Static epistemology has not sufficiently emphasized that the movement and change of reality can be rightly understood only if a historical point of view is adopted.[75]

Likewise, in his textbook titled *Materials and Documents on the History of Physics*, Hessen writes that

> [n]o matter how new and unusual the theories of contemporary physics may be, no matter how radically they differ from the outlook of classical physics, the contemporary stage of development in physics is still a historical phase of its overall development. Therefore, knowledge of the history, of the origin, and development of physical theories not only aids in understanding its contemporary condition, but also helps to establish its historical roots and, by doing so, clears the way for new research.[76]

Additionally, he appeared to reject any notion of a subject-object duality but, instead, articulated a view according to which subject and object form an identity. Thus, "following Mach, [Hessen] rejected any form of dualism between subject and object in the process of knowledge, and believed that the contents of experience do not have an objective meaning."[77] Finally, they contend that Hessen believed that Marxism and relativity theory to be reconcilable theories in principle; this was a position that was more commonly encountered among the Russian Machists than the dialectical materialists. They claim that "Hessen believed, like Bogdanov, that Marxism and relativity were complementary aspects of different visions of the world—the one prioritizing economic aspects of human society and the other physical and cosmic ones."[78]

Second, Ienna and Rispoli argue that Hessen was a Machist to the extent that his thought was deeply indebted to the work of Austrian mathematician and physicist, as well as member of the early Vienna Circle, Richard von Mises.[79] Richard was the younger brother of well-known Austrian School economist, Ludwig von Mises, but Richard, unlike his brother, harbored no strong political convictions and remained a predominantly apolitical intellectual throughout his life. It is no small irony then that Hessen, a figure for whom politics so thoroughly permeated even the natural sciences, was so drawn to such a figure.[80] Nevertheless, from 1928 to 1930, Hessen studied Richard von Mises more closely than any other figure in either mathematics or physics, having written one article, one preface to a translation, one section of an article, and various scattered remarks about him.[81] At one point, Hessen even petitioned the board of the Institute of Red Professors to conduct a research stay with von Mises, who was a professor and appointed director of the Institute of Applied Mathematics at the University of Berlin at the time.[82] No record remains of the institute's reply to Hessen's request, but scholars generally agree that the stay was indeed approved.[83] Von Mises is rightly characterized as an unapologetic defender of Machist epistemology, epitomized perhaps in an article that, while Hessen would never have had the opportunity to read it, nevertheless best summarizes von Mises's position throughout his career. He refers to Mach as "*the most effective critical philosopher of the last generation and the most typical one of our time.*"[84] Precisely what he derives from Mach is his objection to grounding the natural sciences in philosophical materialism but, rather, maintaining that "the world is *given to us only by our sense impressions*."[85] Like Mach, von Mises rejects the concept of the thing in-itself, arguing that "[i]f we take away *all* the properties of an object, what remains is not its 'substance', but nothing—as with an onion, if we go on peeling it."[86] For Ienna and Rispoli then, von Mises's influence betrays the fact that Hessen was a secret adherent to Machist epistemology all along.[87]

Analysis: The Hessen Paradox

The three aforementioned theses offer compelling attempts by which to resolve the discrepancy between Hessen's 1931 Newton paper and the rest of his oeuvre, but I respectfully maintain that they bear certain particular shortcomings, such that they are unable to adequately address the Hessen paradox. First and foremost, the theses rest upon certain assumptions about Hessen's biography that have recently been called into question in the secondary literature. They rest on the belief, for instance, that the Soviet delegation to the London conference was arranged to test the loyalty of delegates like Hessen.

As Sergei Korsakov indicates, however, according to recently discovered minutes of the Organizing Bureau and Secretariat of the Central Committee of the Russian Communist Party, the London trip was not at all arranged under special circumstances but was, instead, organized according to more or less standard bureaucratic procedures.[88] Korsakov also indicates that in his in his recently discovered debriefing report on the London conference, "Speech at the Presidium Session of the Communist Academies," Hessen offers no defense of his talk, choosing instead to focus on strategies by which to cultivate the relationships with British academics who were sympathetic to socialism.[89] C. A. J. Chilvers even states that Hessen's execution may have had less to do with his writings and more to do with his having been a Menshevik internationalist prior to 1917, with Stalin's efforts to manipulate Hessen's wife to help bring down Bukharin.[90]

Second, I maintain that these hypotheses rest on the assumption that Hessen would have had to disguise his adherence to Machist epistemology or favorability to quantum mechanics and relativity theory. This argument, however, not only affords too much faith to the word of Hessen's professional rivals in an environment that was notorious for bad-faith accusations but also does not properly assess the status of Hessen's positions in the Soviet context. For instance, it is not clear that Hessen would have actually been in any grave danger for studying either von Mises or Mach, as thinkers such as Aleksandr Khinchin, Leonid Mandelstam, Mikhail Leontovich, and Andrey Kolmogorov were never touched and all of whom were proponents of von Mises' work, and who devoted substantially less attention to the works of Marx, Engels, and Lenin.[91] Likewise, quantum mechanics and relativity theory were simply not the forbidden topics in the Soviet Union that they are often made out to be in the academic literature. As Alexei Kojevnikov argues, while certain Marxists were indeed vehemently opposed to quantum mechanics and to relativity theory, these were in fact the exception more than the rule. The Communist Party and Marxist intellectuals alike were often favorably disposed to the study of these theories and were less preoccupied with censoring them and more with debating over how these theories could be reconciled with dialectical materialist philosophy.[92]

Third, these three theses do not adequately account for the content of Hessen's 1931 paper and other texts based on their respective audiences. Hessen wrote the 1931 paper for a primarily non-Marxist audience, which would explain the text's more programmatic approach, while virtually everything else he wrote was for a Soviet publication—that is, for an audience who was either already adherent to, or at least familiar, with Marxism and, thus, a philosophical school with whom Hessen shared several, fundamental convictions. Hessen does not explicitly acknowledge this difference anywhere in his work, but it would nevertheless be fair to assume that he would have been more inclined to outline his premises in the former, while this would have been unnecessary in the latter.

Fourth, concerning the Machism thesis specifically, I claim that this position is based on the idea that Hessen shared certain positions with both Alexander Bogdanov and Richard von Mises. The commonalities between Bogdanov's and Hessen's positions, however, are in fact beliefs that were common among most Marxist philosophers. For instance, Hessen's assertion that scientific theories involve an understanding of the historical process from which they culminate can be found not only in the work of Bogdanov but also in the writings of Marx, Engels, and Lenin, as well as in Hessen's immediate predecessor, Abram Deborin.[93] Likewise, in a coauthored article with Ivan Podvolotskiy, Hessen vehemently criticizes Bogdanov precisely for both his Machist epistemology and his opposition to dialectics in favor of equilibrium theory.[94] Regarding von Mises, I maintain that the Machism thesis overlooks precisely what Hessen explicitly claims to be doing in his engagement with his work. Hessen states that he supports von Mises's approach to statistical analysis via the concept of the collective and that he aims to extract this approach from a Machist epistemological framework. Von Mises's adherence to Machism leads to more concrete errors, such as a "metaphysical explanation of the concept of limit and limiting case" and an improper conception of "divisibility and indivisibility."[95] This is consistent with Hessen's overall philosophical project, in which he often adapted or applied dialectical materialistic philosophy to new scientific discoveries.

Fifth, and ultimately, these theses downplay the significance of dialectical materialism and not just in the 1931 piece but also throughout the rest of Hessen's oeuvre. In 1931, Hessen adopted rhetoric that was common in the Soviet discourse of the time, but rather than presenting a "flamboyant" and "elementary" brand of dialectical materialism, he presented one that was quite sophisticated and that was based on careful readings of his predecessors. Moreover, in the Soviet texts, Hessen did not so much write on topics such as quantum mechanics and relativity theory but on the possibility of dialectical materialist accounts of those theories. Additionally, Graham's claim that Hessen omitted socioeconomic explanations from his pre-1931 writings has turned out to be false. For instance, in his recently discovered talk titled "Materialist Dialectics and Modern Physics," Hessen traces changes in the natural sciences to changes in socioeconomic forms on several different occasions.[96] This only appears to become truer the deeper we are able to delve into Hessen's previously untranslated writings.

Nevertheless, each of these three hypotheses rightly pinpoints why Hessen's 1931 Newton paper is so fundamentally vexing in the context of his oeuvre. Hessen spends the entire opening of the 1931 piece criticizing the so-called idealistic predilections of both most historians and philosophers of science, only to apparently adopt that very same approach in most of his own work. In fact, he spends most of his time in his writings discussing philosophical and scientific ideas in light of those genius individuals who first developed them. Like the aforementioned hypotheses, then, this chapter upholds the notion that Hessen's works seem to operate on two fundamentally different sets of premises. However, to reconcile this apparent discrepancy, it will be necessary to advance an altogether different strategy; that is, I will argue that it is necessary to grasp two distinct layers of Hessen's thought.

IDEOLOGICAL STRUGGLE

In this final section, I will explain how the tension in Hessen's oeuvre can be resolved by, once again, deepening our understanding of his Deborinite dialectical materialist approach to the study of natural scientific theory. First, I show that in the 1931 paper, Hessen's primary concern is to show the way in which the conditions of natural scientific theory arise in the context of a social totality. Secondly, I demonstrate that in the 1927 paper, Hessen explains how natural scientific theory is also marked by an ideological conflict between the philosophical persuasions of both materialism and idealism and that understanding this conflict is necessary for understanding how new scientific theories emerge. Finally, I will provide a summative analysis of these two aspects of his theory to explain how Hessen believed dialectical materialism as not offering an ultimate natural scientific theory but a method for negotiating the complexity of how new natural scientific theories are developed.

Practice and Ideology in the 1931 Newton Paper

In the 1931 text, Hessen provides a dialectical materialist account of how natural scientific theory takes shape in the context of the social totality. He bases his approach on Marx's assumption that "[p]ractice should not be explained by ideas, but on the contrary, ideological structures should be explained by material practice."[97] Typically, scholars have taken this to mean that Hessen subscribes to an "economic determinist" model of dialectical materialism.[98] According to this account, the social totality is composed of two parts: first, the "socioeconomic base," that is, the means of production and relations of production, which pertain to the realms of practice; and second, the "superstructure," that is, the "legal, political, religious, artistic or philosophic" structures, which pertain to the realm of ideology.[99] Natural scientific theory, then, would passively derive its content from both realms, thereby making it divisible into two parts.

However, and as mentioned in the previous chapter, Hessen never explicitly defines either practice or ideology in this text, leaving one to study his usages of the terms to sufficiently understand their meaning.[100] As I have argued, Hessen's interpretation of practice can be understood in light of Marxist texts that would have been available to him at the time. For instance, on the one hand, Hessen treats practice as something similar to Marx, Engels, and the Lenin of *Materialism and Empirio-Criticism*, according to whom practice refers to the transformation of the thing in-itself into a thing for-us; that is, practice refers to the transformation and appropriation of objects in the world.[101] On the other hand, though, Hessen treats practice as something similar to Hegel and the Lenin of the *Philosophical Notebooks*, according to whom practice refers to the transformation of the thing in-itself into a thing for-itself; in other words, practice also refers to the process of becoming self-conscious.[102] Taken together, Hessen sees the ultimate goal of dialectical materialism as the actualization of the full potential of human practice. Practice remains inhibited by factors that "naturalize" the present condition, that is, ideology, because practice exists in a deficient mode in class society.

However, Hessen understands practice and ideology to be present in the socioeconomic base and superstructure in a distinct way from the economic determinist model. It would be easy to think that because subtheses 1 and 3 refer to machine technology (i.e., a mean of production), that they also pertain to the realm of practice and that because subtheses 2 and 4 refer to philosophy and theology (i.e., superstructures), that they also pertain to the realm of ideology. Recall, however, that in subthesis 1 that Hessen explicitly states that Newton was exposed to mechanics in the context of the Royal Society; an organization founded by and for the ascendant bourgeoisie of late seventeenth-century England. Theoretical mechanics grew, in part, out of the study of machines in this context; machines were studied for the sake of improving their efficiency in early capitalist manufacture. That is, machines were not studied through the eyes of laborers, but through those of the owners of the means of production. Now, in subthesis 4, when Hessen criticizes machine-wreckers for falling prey to bourgeois "machine fetishism," he means explicitly that the working class remained under the sway of bourgeois ideology. So, while the study of machines in the Royal Society and machine-wrecking pertain primarily to the realm of practice, they are both colored by the ideology that singled out machinery as the object of study/protest in the first place. Conversely, we should note that in subthesis 2 that philosophy and theology do not simply exist in the minds of their believers for Hessen but were actually vehicles for advancing concrete forms of practice. Back in Newton's day, materialism was the philosophical persuasion that was advanced by absolutist monarchists (i.e., Hobbes) or by radical democrats (i.e., Overton), while idealism was favored by those among the bourgeois/merchant class who sought constitutional monarchy and to share parliamentary power with the nobility.[103]

For Hessen, then, both practice and ideology cannot be so easily attributed to the socioeconomic base and the superstructure, respectively, as in the economic-determinist model. He maintains that practice and ideology will constitute a unity of opposites as long as class society still exists; that is, while the socioeconomic base primarily pertains to the realm of practice, it is also infused with ideology, and while the superstructure primarily pertains to the realm of ideology, it is also infused with practice.[104] A natural scientific theory uncorrupted by ideology, then, would only be possible in the context of a society marked by fully actualized practice—that is, in a classless society. That condition notwithstanding, however, natural scientific theory reflects the unity in opposition of practice and ideology, meaning that it must recognize that one can derive practical elements from the superstructure and must be on guard against ideological elements from the base. This, I believe, is the real point of contention between Hessen's brand of dialectical materialism and Stalin's diamat. For Hessen, Stalin's declaration that "technology in the current stage decides everything" was simply another form of machine fetishism, an embrace of blind ideology at the expense of practice. Thus, natural scientific research itself is a sphere animated from within by a unity in opposition of practice and ideology but also because of the distribution of practice and ideology found throughout the social totality.

Idealism and Materialism in the *Under the Banner of Marxism* Papers

As indicated previously, Hessen's 1927 paper is one among several in an ongoing series of articles in the journal, *Under the Banner of Marxism*. From 1927 to 1930, he contributed nine papers to the journal overall, five of which are coauthored. Apart from his 1931 Newton paper, his contributions to *Under the Banner of Marxism* stand as the other, clearest articulation of his methodology towards the history and philosophy of science. To provide some background, *Under the Banner of Marxism* ran from January 1922 to June 1944 and was published monthly save for between 1933 and 1935, when it was issued only every other month. The successor to the journal, *Problems of Philosophy*, was established in July 1947. Over the course of its twenty-two-year lifespan, *Under the Banner of Marxism* was considered to be one of the most important platforms for the elaboration of the principles of

Marxism in the Soviet Union, particularly in the late 1920s and early 1930s, the period during which Hessen was most active.[105] The journal's significance is due in no small part to mandates by Leon Trotsky and V. I. Lenin over its overall purpose. In his "Attention to Theory: Letter to the Editor of *Under the Banner of Marxism*" from February 27, 1922, Trotsky declared that the journal could serve as "a magazine that would introduce advanced proletarian youth into the circle of materialist ideology."[106] In his "On the Significance of Militant Materialism" from March 22, 1922, Lenin also argued that the primary function of the journal was to operate as a "Society for Materialist Friends of the Hegelian Dialectic."[107] Lenin based this declaration on the idea that revolutionaries often made the mistake of thinking that they alone could make the revolution happen; he insisted, however, that the revolution could not continue without the cooperation of nonrevolutionaries. He argued that this was most urgent with respect to garnering the support of natural scientists, engineers, and so on, for their practical and theoretical expertise.[108] In cultivating this cooperation, a twofold responsibility fell to Marxist theoreticians: first, to study the relationship between Hegelian and Marxist dialectics, and second, to study the ensuing revolution in early twentieth-century physics. Theoreticians were expected to study the relationship between Hegelian and Marxist dialectics to understand how Hegelianism, considered to be the culmination of bourgeois theory, contained the potential to become Marxism, considered to be the inauguration of proletarian theory. Lenin believed that strategies could be developed for converting natural scientists, engineers, and others into defenders of the spirits of the October Revolution by better understanding the passage from the former to the latter. This engagement was an additional imperative as Marxists could no longer rely on classical mechanics for practical and theoretical purposes and depended on the knowledge of natural scientists, engineers, and others, to develop a better understanding of natural and applied sciences for the future. For Lenin, *Under the Banner of Marxism* marked the perfect venue for staging discussions on such matters.[109]

The papers in the journal are of a highly polemical nature, one article often being a response to another, and Hessen's papers are no exception; this is because there was no single, agreed-upon approach about how to take up these challenges. Most of his writings are interventions in the ongoing debate between the mechanists and Deborinites/dialecticians—the mechanists favoring classical mechanics and the dialecticians favoring contemporary physical theories—over what constituted the properly dialectical materialist approach to the natural sciences. Hessen also wrote summaries of notable works by contemporary physicists, which also served as interventions in the ongoing debate with the mechanists. He embraces a swath of different themes throughout these papers, but the common thread that runs through each of them is his dialectical materialist understanding of the significance of ideology in the development of natural scientific theory. According to Hessen, the "dialectical" in dialectical materialism does not mean unqualifiedly embracing any brand of materialism or, conversely, unqualifiedly dismissing any brand of idealism but, instead, involves the recognition that idealism and materialism form their own unity of opposites within the history of the development of natural scientific theory. Materialism takes precedent over idealism within this unity, but there are moments in which idealism takes a principal role. These mark important transitional periods that signal the breakdown of an old paradigm of materialism and the emergence of a new one.

Let us carefully retrace Hessen's reasoning in support of this position. Like the mechanistic materialists, Hessen believed that the philosophical question over idealism or materialism had been sufficiently arbitrated by Marx, Engels, and Lenin. Lenin, in particular, had dealt a decisive blow to idealism with his contention in *Materialism and Empirio-Criticism* that only materialists could account for the traces of mind-independent reality, namely, those of prehistoric life in the fossil record.[110] However, there is an important distinction, according to Hessen, between "a purely physical and a philosophical concept of matter."[111] The philosophical concept of matter refers to the mere notion that mind-independent reality exists. The physical concept, though, refers to the manifold ways in which that mind-independent reality can manifest, that is, composition, forms of motion, and so on. So, while the philosophical concept of matter may be true, it leaves some of the most important questions about matter unanswered; these are questions that can only be dealt with through ongoing revision. Hessen notes that Engels and Lenin themselves defend this claim, as they argue that "'with each epoch-making discovery even in the sphere of natural science ['not to speak of the history of mankind'], materialism has to change its form."[112] Hessen further indicates that in these changes of form throughout the history of the natural sciences, idealistic theories have continually played the role of signaling the breakdown of an old paradigm of materialism and the emergence of a new kind. Citing *Materialism and Empirio-Criticism* again, he notes that although physicists such as Duhem, Stallo, and Rey advanced idealistic, natural scientific theories, they

> most energetically combat[ed] the atomistic-*mechanical* conception of nature. They show[ed] the narrowness of this conception, the impossibility of accepting it as the limit of our knowledge, the *rigidity* of many of the ideas of writers who hold this conception. And it is undeniable that the *old* materialism did suffer from such a defect.[113]

Hessen, in one of the articles he cowrote with Vasiliy Egorshin, provides several further examples, noting that "[a]t one time, Balmer's formula was also just a description, but was it not also a step forward? Bohr came along after Balmer and explained the physical significance of the formula,"[114] and that "[a]t one time, idealism used for its own purposes an electron theory by claiming that it contradicts materialism," but that eventually "electron theory turned out to be, not at all a refutation, but a brilliant confirmation of dialectical materialism."[115] Hessen argues that this recurring theme throughout history indicates that idealism and materialism coexist in a more complex relationship than mechanistic materialists had ever assumed.[116]

There is further precedent for this position in the Marxist tradition, Hessen maintains, in Lenin's more abstract philosophical analyses in "On the Question of Dialectics," in which he argues that from the perspective of dialectical materialism, idealism is an important moment in the process of the development of knowledge, which can, on the one hand, become exaggerated and tangential, but, on the other hand, it also plays a necessary role in the developmental process. He writes that "[p]hilosophical idealism is *only* nonsense from the standpoint of crude, simple, metaphysical materialism. From the standpoint of dialectical materialism, on the other hand, philosophical idealism is a *one-sided*, exaggerated, überschwengliches (Dietzgen) development (inflation, distension) of one of the features, aspects, facets of knowledge into an absolute."[117] In other words, the idealistic digression can bear one of two outcomes: either the slippage into mysticism, solipsism, and so on, or the enrichment of materialism. For Hessen, the dialectical standpoint is the only point of view that properly comprehends the momentary deference to idealism. Here again, he cites Lenin's claim that "[t]he basic materialist spirit of physics, as of all modern science, will overcome all crises, but only by the indispensable replacement of metaphysical materialism by dialectical materialism."[118]

For Hessen, this very phenomenon was taking place within the contemporary context. In his brief preface to the Russian translation of Gustav Mie's *The Problem of Matter*,[119] Hessen contends that amidst the crisis of contemporary physics, many physicists, like Mie, had not abandoned materialism for idealism but were, instead, entertaining it as a pathway to a new form of materialism. He notes that while "[n]ew experimental data and the development of theoretical physics are making clear the limits and the unacceptability of the mechanistic worldview to substantiate modern physics."[120] For Mie, the contemporary conflicts illustrated the need to broaden the meaning of the concept of motion from "a simple change of place in space" to "the concept of changing states."[121] In this respect, Hessen identifies Mie as a physicist who is "consciously materialist,"[122] but who remained at risk of drifting into idealistic claims because of his lack of a robust notion of dialectics.[123] According to Hessen, dialectical materialism provided the method for both explaining idealistic digressions as well as for how to anticipate new materialistic explanations.

Indeed, he maintained that Soviet science was doomed to failure if it simply excluded the discoveries of quantum mechanics and relativity theory on the grounds that they flirted with idealism. While both theories could very well turn out to be disproven, they had demonstrated that there was simply no longer any way of retreating to classical mechanics.[124] New scientific discoveries had been rapidly eroding the ground beneath classical mechanics, such that if Marxists insisted on treating it alone as the only possible materialism, they risked locking themselves into ever more absurd positions.[125] They argue that rather than simply dispensing with new theories, "[i]f, at present, Einstein's theory is not yet rigorously, experimentally proven, then Marxist methodology must await the resolution of the physical question by physical methods and cut off any . . . attempts to use this question . . . for the benefit of Machist, fictional or any other idealistic philosophy."[126] To do otherwise, they insist, would be "*blatant nihilism.*"[127] He and Egorshin proceeded to adamantly uphold a cooperative venture between the Soviet state and contemporary natural scientists:

> [w]e, together with the Party and Soviet power, adhere to the unshakable opinion that we need to work together with the representatives of modern science, that we can make them dialectical materialists only by collaborating with them, and we maintain that the nihilistic standpoint would bring irreparable harm to Marxism and to the revolution.[128]

Analysis: Ideological Struggle

From these two accounts, Hessen does not appear to articulate two distinct methodologies at all but, rather, two different aspects of one and the same methodology. As indicated previously, practice for him refers to the interrelated processes of transforming things in-themselves into things for-us as well as transforming things in-themselves into things for-themselves; that is, practice is the interrelated effort of transforming and appropriating nature and of acquiring self-consciousness. Ideology refers to that which obscures the transient nature of class society, a blind spot that renders the contemporary situation as natural. However, practice remains locked into a unity of opposites with ideology as long as class society persists. So long as this unity of opposites persists, practice cannot be located solely in the socioeconomic base of society, just as ideology cannot be located solely in the superstructure; practice and ideology remain diffuse in various forms throughout the entirety of society. This means that natural scientific theory not only occupies an ambiguous space between the socioeconomic base and superstructure but also that natural scientific theory remains animated by practice and ideology from within. For this reason, while materialist natural scientific theories uphold the superior paradigm, idealistic deviations still play an important role in the development of natural scientific concepts. Moreover, Hessen discourages the notion that dialectical materialism will ever offer a single, monolithic theory of the natural world. As he argues, "[s]cience and scientific truth are not something absolute, given once and for all. 'Science is a creative movement of knowledge.' The change of scientific truths is a necessary stage in the process of the unlimited approximation to absolute truth, reflecting change in socioeconomic forms."[129] In other words, dialectical materialism is a method meant for the negotiation of the ongoing development of natural scientific theories from now unto infinity.

NOTES

1. This chapter is a revised version of my "The Materialist Dialectic in Boris Hessen's Newton Papers (1927 and 1931)," *Historical Materialism* 28.4 (2020): 202–34, and also consists of passages from my "Individuals and Collectives in the Philosophy of Boris Hessen: An Introduction," *Science in Context* 34.1 (2021): 121–36.
2. Schäfer, "Boris Hessen and the Politics of the Sociology of Science," 113.
3. Graham, "The Socio-Political Roots of Boris Hessen," 713.
4. J. B. Morrell, "Externalism," in *Dictionary of the History of Science*, ed. W.F. Bynum, E. J. Browne and R. Porter (London: The Macmillan Press Ltd., 1981), 145. See also Graham, "The Socio-Political Roots of Boris Hessen," 706; Schäfer, "Boris Hessen and the Politics of the Sociology of Science," 103.
5. J.B . Morrell, "Internalism," in *Dictionary of the History of Science*, ed. W.F. Bynum, E. J. Browne and R. Porter (London: The Macmillan Press Ltd., 1981), 211.
6. See chapter 1.
7. Hessen, "The Social and Economic Roots of Newton's *Principia*," 44, 61.
8. Ibid., 43–61.
9. Ibid., 61–73.
10. Ibid., 73–82.
11. Ibid., 82–89.
12. А. Эйнштейн, «Механика Ньютона и ее влияние на развитие теоретической физики», *Под знаменем марксизма* 4 (1927): 166–73. See also A. Einstein, "The Mechanics of Newton and Their Influence on the Development of Theoretical Physics," in *Ideas and Opinions*, ed. C. Seeling, et al. and trans. S. Bargmann (New York: Crown Publishers, 1954 [1927]), 253–61.

13. Дж. Дж. Томсон, «Работы Ньютона в области физики», *Под знаменем марксизма* 4 (1927): 174–81. See also J. J. Thomson, "Newton's Work in Physics," Nature 119 (1927): 36–40.

14. Г. Ламб, «Работы Ньютона в области механики», *Под знаменем марксизма* 4 (1927): 182–85. See also H. Lamb, "Newton's Work in Mechanics," *Nature* 119 (1927): 33–35.

15. B. Hessen, "Preface to Articles by A. Einstein and J.J. Thomson," trans. S. Winkler, *Society and Politics – Boris Hessen and the Dialectics of Natural Science* 13.1 (2019 [1927]): 87.

16. Ibid., 88.

17. Ibid., 94.

18. Ibid., 100.

19. Ibid. See also Schäfer, "Boris Hessen and the Politics of the Sociology of Science," 114.

20. Hessen, "Preface to Articles by A. Einstein and J.J. Thomson," 91.

21. E. Halley, "Halley to Newton (29 June 1686)," in *The Correspondence of Isaac Newton, vol. 2 – 1676–1687*, ed. H. W. Turnbull, F.R.S. (Cambridge: Cambridge University Press, 1960 [1686]), 441. See also Hessen, "Preface to Articles by A. Einstein and J.J. Thomson," 93.

22. Hessen, "Preface to Articles by A. Einstein and J. J. Thomson," 93.

23. Ibid.

24. G. Plekhanov, *The Role of the Individual in History*, trans. unknown (New York: International Publishers, 1940 [1898]), 11n1. See also Hessen, "Preface to Articles by A. Einstein and J.J. Thomson," 93.

25. Hessen, "Preface to Articles by A. Einstein and J. J. Thomson," 93.

26. Ibid., 92.

27. Ibid., 89.

28. Ibid.

29. Ibid., 92.

30. Ibid., 91.

31. Ibid., 92.

32. Ibid., 90.

33. Ibid., 89.

34. Ibid., 90.

35. Ibid., 92.

36. Ibid., 90.

37. Ibid., 91.

38. Ibid., 94.

39. R. Descartes, *Principles of Philosophy*, in *The Philosophical Writings of Descartes, vol. 1*, ed. and trans. J. Cottingham, R. Stoothoff and D. Murdoch (Cambridge: Cambridge University Press, 1985 [1644]), 247. See also Hessen, "Preface to Articles by A. Einstein and J.J. Thomson," 89.

40. Hessen, "Preface to Articles by A. Einstein and J. J. Thomson," 94.

41. Graham, "The Socio-Political Roots of Boris Hessen," 718.

42. Ibid., 716.

43. Ibid., 705–7.

44. J. Stalin, "A Year of Great Change: On the Occasion of the 12th Anniversary of the October Revolution," in *J. Stalin: Works, vol. 12 – April 1929–June 1930*, ed. and trans. Marx-Engels-Lenin Institute (Moscow: Foreign Languages Publishing House, 1954 [1929]), 124–41.

45. Graham, "The Socio-Political Roots of Boris Hessen," 707.

46. Ibid., 715–16.

47. Ibid., 713, 714.

48. Ibid., 713.

49. Ibid., 714–16.

50. Joravsky, *Soviet Marxism and Natural Science, 1917–1932*, 185–86; Graham, "The Socio-Political Roots of Boris Hessen," 712; Schäfer, "Boris Hessen and the Politics of the Sociology of Science," 114.

51. Joravsky, *Soviet Marxism and Natural Science*, 286; Graham, "The Socio-Political Roots of Boris Hessen," 712.

52. Joravsky, *Soviet Marxism and Natural Science*, 187; Graham, "The Socio-Political Roots of Boris Hessen," 711, 712; Schäfer, "Boris Hessen and the Politics of the Sociology of Science," 115.

53. Graham, "The Socio-Political Roots of Boris Hessen," 713.

54. Ibid.

55. Ibid., 714–15.

56. Ibid., 707.

57. Ibid., 716.

58. Ibid., 715–16.

59. Ibid., 717.

60. D. Joravsky, "Soviet Views on the History of Science," *Isis* 46.1 (1955): 3–13. See also Schäfer, "Boris Hessen and the Politics of the Sociology of Science," 113.

61. Ibid., 114.

62. Admittedly, at one point in his paper, Graham writes that "[b]y placing emphasis on technology and practice in determining theoretical physics he freed physics itself from being condemned merely by the philosophical or theological interpretations that may be placed on it. He believed that the development of twentieth-century physics could be analysed in the same way that he explained Newtonian physics, and thought that there was no more reason to accept attacks on materialism in the name of twentieth-century physics than there had been to accept such attacks in the name of Newton, whose religious views were merely a 'product of his time and class'." This suggests that for Graham, Hessen makes a distinction between those external factors that contribute to the development of the natural sciences (i.e., technology and practice), and those that impede its development (i.e., philosophy or theology) because they are contingent upon time and class. However, Hessen argues that Newton's time and class are also at play in both his understanding of technology and practice, insofar as the study of actual machines is informed by the rising bourgeoisie's need to improve upon machines in the realm of economic production. Moreover, Hessen does not attribute a strictly reactionary character to his system's philosophical and theological aspects, given that he considers them to have been progressive in their own

day. For an elaboration on this claim, see The Ideological Structure in this chapter. Graham, "The Socio-Political Roots of Boris Hessen," 717.

63. Ольга Столярова, «Идеи Б.М. Гессена и отечественная философия», *Философский журнал* 10 (2017): 121.

64. Ibid., 123.

65. Ibid.

66. Ibid.

67. Ibid., 125.

68. Ibid., 128.

69. Ibid., 122–25, 127–28.

70. Ienna and Rispoli, "Boris Hessen at the Crossroads of Science and Ideology," 52–53.

71. Ibid., 58.

72. Ibid., 54.

73. Ibid., 56.

74. Ibid., 53–54.

75. А. А. Борданов, *Познание с исторической точки зрения* (Санкт-Петербург, 1899), 10–19. See also Ienna and Rispoli, "Boris Hessen at the Crossroads of Science and Ideology," 55–56.

76. Boris Hessen, *Manuscripts and Documents on the History of Physics: A Historical Materialist Textbook*, ed. Pietro Omodeo and Sean Winkler and transl. Giuliano Vivaldi (Venice: Verum Factum, 2022), 187. See also Ienna and Rispoli, "Boris Hessen at the Crossroads of Science and Ideology," 56.

77. Ienna and Rispoli, "Boris Hessen at the Crossroads of Science and Ideology," 53–54.

78. Ibid., 54.

79. Ibid., 53.

80. Ibid.

81. See B. Hessen, "Selected Material on the Work of Richard von Mises," in *Boris Hessen: Physics and Philosophy in the Soviet Union, 1927–1931; Neglected Debates on Emergence and Reduction*, ed. and trans. C. Talbot and O. Pattison (Cham, Switzerland: Springer Nature, 2021), 131–39.

82. Hessen's petition reads as follows: "'[p]lease provide me with a business trip to Germany. Currently, I am working on the problem of substantiating statistical mechanics and applying the statistical method in physics. I work under the supervision of Professor L.I. Mandelstam. In the mathematical and methodological part, the work represents the development of the ideas of Richard Mises. Mises' works have appeared in recent years and so far are very little known. To complete the work, I would need to work with Mises, who is a professor at the University of Berlin. Mises is involved in working with Mandelstam, whom I currently work for, and so Mises will have favorable working conditions. In the summer and fall of this year in Berlin, Einstein and Planck organize special courses for foreign physicists (*Für Ausländer*). The best European physicists will read at these courses: Planck, Einstein, Ehrenfest, Mises, etc. This case is unique in its kind, because such courses are not held every year. Mises will also take part in the courses, giving the course just on the topic that

interests me. Einstein will also read about causality and statistics. In general, courses have a methodological bias, which is especially valuable. This year I was busy with a lot of organizational work both inside the Institute (deputy head of the natural department) and outside it. Therefore, I would like to use the remaining months especially intensively to complete my first serious independent work. I agree to share the full business trip in four months in half with Egorshin. Due to the fact that I'm teaching at the Physics and Mathematics Faculty of Moscow State University, I ask you to provide a business trip from May 15 to September 15.'" Корсаков, Козенко, and Грачева, Борис Михайлович Гессен, 19–20. See also Ienna and Rispoli, "Boris Hessen at the Crossroads of Science and Ideology," 50; A. Pechenkin, *L.I. Mandelstam and His School in Physics*, 2nd ed. (Basel: Springer Natural Switzerland AG, 2019), 97.

83. Pechenkin, *L. I. Mandelstam and His School in Physics*, 97.

84. R. von Mises, "Ernst Mach and the Scientific Conception of the World," in *Unified Science: The Vienna Circle Monograph Series originally edited by Otto Neurath, now in an English Edition*, ed. B. McGuinness and trans. H. Kaal (Dordrecht: D. Reidel Publishing Company, 1987 [1938]), 170.

85. Ibid., 178, 187.

86. Ibid., 183.

87. Ienna and Rispoli, "Boris Hessen at the Crossroads of Science and Ideology," 53.

88. С. Н. Корсаков, «Слово товарищу гессену», *Эпистемология и философия науки* 3 (2018): 203.

89. Ibid. See also Б. М. Гессен, «Выступление на заседании президиума коммунистической академии», ed. С.Н. Корсаков, *Эпистемология и философия науки* 3 (2018 [1931]): 205–10.

90. Chilvers, "The Tragedy of Comrade Hessen: Biography as Historical Discourse," 111–15.

91. L. M. Verburgt, "'A Terrible Piece of Bad Metaphysics'?: Towards a History of Abstraction in 19th- and early 20th-Century Probability Theory, Mathematics and Logic" (PhD diss., University of Amsterdam, 2015), 562–63.

92. A. Kojevnikov, "The Phenomenon of Soviet Science," *Osiris* 23 (2008): 126–27; A. Kojevnikov, "Probability, Marxism, and Quantum Mechanics," *Yearbook for European Culture of Science* 6 (2011): 213.

93. For example, in *Hegel and Dialectical Materialism*, Deborin writes as follows: "It is wrong to think that the history of philosophy for more than two and a half thousand years was a fruitless struggle of various opinions about worthless things and that this struggle has now ended in nothing, a round zero. People who are not familiar with the history of philosophy and science usually think so. . . . Modern giant successes in the field of natural science and technology would be impossible and inconceivable without all the previous history of mankind. . . . [T]he same thing has to be said about philosophy. Moreover, philosophy and science have always been so closely interconnected that they mutually feed each other". A. Deborin, *Hegel and Dialectical Materialism*, trans. unknown (Online, 1929), www.marxists.org.

94. Гессен and Подволоцкий, «Философские корни правого оппортунизма», 1–29.

95. Hessen, "Selected Material on the Work of Richard von Mises," 137.

96. B. Hessen, "Materialist Dialectics and Modern Physics: Abstracts of the Report at the First All-Union Congress of Physicists in Odessa on 19 August 1930," trans. S. Winkler, *Historical Materialism* 28.4 (2020 [1930]): 236, 237.

97. Hessen, "The Social and Economic Roots of Newton's *Principia*," 43.

98. R. Young, "Marxism and the History of Science," in Companion to the History of Modern Science, ed. R. C. Olby, et al. (London: Routledge, 1990), 77–79.

99. Marx and Engels, *The German Ideology*, 59; Marx, *A Contribution to the Critique of Political Economy*, 263. See also Hessen, "The Social and Economic Roots of Newton's *Principia*," 43, 87, 96n6, 101n86.

100. See first section in this chapter.

101. Lenin, *Materialism and Empirio-Criticism*, 102–3, 122.

102. Lenin, "Conspectus of Hegel's Book *The Science of Logic*," 211, 212, 218.

103. Hessen, "The Social and Economic Roots of Newton's *Principia*," 62–67.

104. See L. Althusser, "Ideology and Ideological State Apparatuses (Notes towards an Investigation)," in *Lenin and Philosophy and Other Essays*, trans. B. Brewster (New York: Monthly Review Press, 1971 [1970]), 170.

105. See A. A. Ермичёв, Рецензия на *Журнал* Под знаменем марксизма *1922–1944. Указатель содержания, Вопросы философии* 7 (2012): 184–86.

106. L. Trotsky, "Attention to Theory: Letter to the Editor of *Under the Banner of Marxism*," trans. Y. Garmize and A. Gonopolskiy, *Platypus Review* 34 (2011 [1922]): 1.

107. Lenin, "On the Significance of Militant Materialism," 233–34.

108. Ibid., 227, 232.

109. Ibid., 233–34.

110. Lenin, *Materialism and Empirio-Criticism*, 72–77, 80–85, 115, 123, 132.

111. B. Hessen and V. Egorshin, "The Fifth Congress of Russian Physicists," in *Boris Hessen: Physics and Philosophy in the Soviet Union, 1927–1931; Neglected Debates on Emergence and Reduction*, ed. and trans. C. Talbot and O. Pattison (Cham: Springer Nature, 2021 [1927]), 17.

112. Engels, *Ludwig Feuerbach and the End of Classical German Philosophy*, 369–70. See also Lenin, *Materialism and Empirio-Criticism*, 251; B. M. Hessen and V. P. Egorshin, "On Cde. Timiryazev's Attitude towards Contemporary Science," trans. S. Winkler and rev. A. Kojevnikov, *Science in Context* 34.1 (2021): 148.

113. Lenin, *Materialism and Empirio-Criticism*, 316. See also Hessen and Egorshin, "On Cde. Timiryazev's Attitude towards Contemporary Science," 145.

114. Lenin, *Materialism and Empirio-Criticism*, 316. See also Hessen and Egorshin, "On Cde. Timiryazev's Attitude towards Contemporary Science," 146.

115. Lenin, *Materialism and Empirio-Criticism*, 316. See also Hessen and Egorshin, "On Cde. Timiryazev's Attitude towards Contemporary Science," 148.

116. Lenin, *Materialism and Empirio-Criticism*, 316. See also Hessen and Egorshin, "On Cde. Timiryazev's Attitude towards Contemporary Science," 148.

117. V. I. Lenin, "On the Question of Dialectics," in *V.I. Lenin: Collected Works, vol. 38; Philosophical Notebooks*, ed. S. Smith and trans. C. Dutt (Moscow: Progress Publishers, 1972 [1925]), 361.

118. Lenin, *Materialism and Empirio-Criticism*, 312. See also Hessen and Egorshin, "On Cde. Timiryazev's Attitude towards Contemporary Science," 145.

119. Г. Ми, «Проблема материи», *Под знаменем марксизма* 1 (1927 [1925]): 122–33. See also G. Mie, *Das Problem der Materie; öffentliche Antrittsrede gehalten am 26 Januar 1925* (Freiburg in Baden: Speyer and Kearner, 1925).

120. Б. М. Гессен, «Предисловие к статья Г. Мие «Проблема материи»», *Под знаменем марксизма* (1927): 118.

121. Ibid.

122. Ibid., 119.

123. Ibid.

124. Hessen and Egorshin, "On Cde. Timiryazev's Attitude towards Contemporary Science," 148.

125. Ibid.

126. Ibid.

127. Ibid, 145.

128. Ibid, 152.

129. Hessen, "Materialist Dialectics and Modern Physics," 236.

Chapter Three

Modern Physics

In this third chapter,[1] I will proceed by showing how Boris Hessen went about applying his Deborinite dialectical materialist approach to the predominant theories of modern physics (i.e., quantum mechanics and relativity theory). Philosophers and scientists were not altogether clear about where they stood in relation to these two revolutionary positions in the Soviet Union of the late 1920s. Soviet theorists were sometimes skeptical of the extent to which they were compatible with their political project, given that both had been developed by natural scientists in so-called "bourgeois" contexts and because many of the proponents of these theories embraced Machist epistemology. Those dialectical materialists of the mechanist persuasion specifically saw quantum mechanics and relativity theory as being incompatible with the aims of a workers' state, precisely on the aforementioned grounds, while those dialectical materialists of the Deborinite persuasion, Hessen included, saw the wholesale rejection of these theories as crude and simplistic and, ultimately, as threatening to stunt the growth of the Soviet project.[2] Recall that for Hessen in particular, the history of science betrays both cyclical and developmental patterns; these were patterns that would be perilous to ignore if the nation were to keep pace with scientific progress elsewhere in the world.[3] As argued previously, Hessen maintains that because new scientific theories emerge from the discovery of new data that are incompatible with established conceptions of materialism, they often first appear not as materialist but, rather, as empiricist or idealist positions.[4] The history of science shows time and again, however, that these more often than not become the germs of altogether new paradigms of materialism. Thus, Hessen insisted that it would be fallacious to simply dispense with the wealth of scientific data and theorizing afforded by the new sciences.

This chapter proceeds, then, by arguing that Hessen saw quantum mechanics and relativity theory as compatible with dialectical materialism to the extent that they bought sought to usher in entirely new ways to understand materialism as such. He demonstrates this by maintaining that neither theory can be grasped independently of their existence as moments in the historical development of conceptions of both matter and motion and that their study would be advantageous in the context of a workers' state. The first section, The Struggle for Scientific Socialism, explains how quantum mechanics and relativity theory fit into the broader debate about what exactly constituted 'scientific socialism' in the Soviet context. Then, in sections, Quantum Mechanics and Relativity Theory, I will chart Hessen's argument for how quantum mechanics and relativity theory are compatible with dialectical materialist philosophy and socialist construction. In support of my claims, I will draw from Hessen's oeuvre, paying special attention to his Preface to the Russian translation of Austrian physicist, Arthur Haas's *Wave Mechanics and the New Quantum Theory*, *The Main Ideas of the Theory of Relativity*, and to his contributions to the *Soviet Encyclopedia*, along with selections from the secondary literature.

THE STRUGGLE FOR SCIENTIFIC SOCIALISM

In this first section, I will lay the groundwork for understanding how the theories of quantum mechanics and relativity theory came to be controversial in the Soviet context, due to the ongoing ambiguity regarding the question of what it meant for Marxist theory to be "scientific." From its inception, Marxist theory was inextricably linked with the history and philosophy of science. Neither Karl Marx nor Friedrich Engels ever claimed to be the forefathers of either socialism or communism, but they attributed their originality to the way in which they sought to bring about such a vision of society. While so-called "utopian socialists" based their political programs on the design and establishment of model societies, Marx and Engels claimed to uphold "scientific socialism," according to which history provides no precedent for new societies springing ready-made from the minds of visionaries but, rather, from struggles for power over the productive forces of society. If a socialist or communist society were to come about, they maintained, it would do so in the same way as the feudal and capitalist societies that had preceded had; that is, from the shell of the previous system.[5] However, if Marx and Engels claimed that their socialism was scientific, this nevertheless raised the broader question of what it meant to be scientific in the first place. Throughout the course of their oeuvre, Marx and Engels both announced that their positions were the culmination of the history of philosophy and science, but they used several names for their own scientific position: "historical materialism," "modern materialism," "the new materialism," "the dialectical nature of modern materialism," and so on. Ironically, the term that came to be most closely associated with them (i.e., "dialectical materialism") was not a term that they ever actually used in their writings together.[6] Instead, the term was coined by a leather tanner named Joseph Dietzgen, who developed a highly similar philosophy to Marx and Engels and who even corresponded with them on occasion.[7] The term was later adopted by the so-called "father of Russian Marxism," Georgi Plekhanov and eventually became the basis of the leader of the October Revolution, V. I. Lenin's own dialectical materialist reading of Marxism.[8] To understand exactly what it meant for Marxism to be scientific socialism in the Soviet Union, then, it is first necessary to understand Lenin's dialectical materialism and how his statements about this philosophy in his work, *Materialism and Empirio-Criticism*, took hold. Written between February and October 1908, Lenin first published the text under the pseudonym V. Il'in in May 1909 with Krumbügel's Zveno Publishing House in Moscow, which produced a mere two thousand copies.[9] The text would not see a more significant impact until its rerelease by Lenin in 1920 and upon Lenin's death when it eventually "became an obligatory text studied in all institutions of higher learning in the Soviet Union."[10] The origins of *Materialism and Empirio-Criticism* are essential to understanding the parameters of the debate about the nature of Marxism as scientific socialism eventually took shape throughout the Soviet Union.

To that end, we must turn back time to more than a decade before the October Revolution. The Russian Social-Democratic Labor Party (RSDLP) was in a state of fundamental disarray in the aftermath of the failed Russian Revolution of 1905. While the revolution did lead to a number of significant reforms (i.e., Tsar Nicholas II's concession to form a Duma), it was ultimately a failure, particularly for the RSDLP who saw the revolution as the opening of radical possibilities, only to have them violently suppressed.[11] With the party already divided between Bolshevik and Menshevik factions, 1905 only multiplied subfaction upon subfaction, threatening to collapse intraparty debate under its own weight. Among the most influential figures within the party were the founding members of the Bolshevik faction, Bogdanov and Lenin, in whom two different paths to socialist science were set out. Bogdanov set out to charter a new philosophical foundation for Marxist revolutionary theory in his *Empirio-monism* while serving a prison sentence.[12] Upon reading the text, Lenin was immediately vexed and found himself in a flurry, filled three notebooks with philosophical reflections in a work which came to be known as "Notes of an Ordinary Marxist on Philosophy," and which has since been lost.[13] In 1908, the poet, Maxim Gorky, a mutual acquaintance of both Bogdanov and Lenin and fellow Bolshevik, hosted a meeting between the two men in Capri, Italy, with the hopes of bringing the feud to resolution. However, the encounter only left Lenin convinced that Bogdanov's position was not only philosophically dubious but potentially disastrous to the RSDLP's entire strategy.[14] In the aftermath of the meeting, Bogdanov went on to assemble an edited volume, *Studies in Marxist Philosophy*, with contributions from fellow empirio-critics, such as Vladimir Bazarov, Alexander Bogdanov, Anatoly Lunacharsky, Jakov Berman, Osip Gelfond, Pavel Yushkevich, and Sergey Suvorov.[15] Lenin would respond in due course with a series of his own articles—"Marxism and Revisionism," "An Evaluation of the Russian Revolution," "The Agrarian Question in Russia toward the End of the 19th Century," "Leo Tolstoy as a Mirror of the Russian Revolution," and "The Agrarian Programme of Social Democracy in the Russian Revolution"—all culminating in *Materialism and Empirio-Criticism*.[16]

For Bogdanov, the 1905 Revolution did not call for an abandonment of Marxism, but it did demand a fundamental reevaluation of the foundations of Marxist theory. As he writes, the "old formulation of historical monism, without ceasing to be basically true, no longer fully satisfies us."[17] In the place of the, what he took to be, metaphysical archaisms of dialectical materialism, Bogdanov turned once again to what he considered to be the most highly developed philosophy of science of the day, namely, "Machism" and its descendent "empirio-criticism;" these philosophies of science attributable to Ernst Mach and Richard Avenarius, respectively.[18] The shared insight of these philosophies is, fundamentally, that "complexes of sensation," or experience, are primary to conceptions of body or mind. Mach stated this principle as "the world consists only of our sensations. In which case we have knowledge only of sensations," while Avenarius held that "'[w]e have recognized that the existing . . . is substance endowed with sensation; substance falls away . . . , sensation remains; we must then regard the existing as sensation, at the basis of which there is nothing which does not possess sensation.'"[19] Though Bogdanov did not consider himself to be either a Machist or an empirio-critic, strictly speaking, he did acknowledge that his own "empirio-monism" was heavily indebted to the basic insights of these two philosophies.[20] Like Mach and Avenarius, Bogdanov maintains that neither "bodies" nor "minds" constitute the proper starting point for proper natural scientific investigation. Experience communicates that we only encounter ephemeral qualities, without any contact with the underlying substrate of "things in-themselves." We may conclude that body and mind are nothing more than practical constructs, which dissolve upon deeper theoretical speculation, because we never actually encounter such things. As with the peeling away the layers of an onion, stripping something of its series of profiles reveals, in fact, nothing underneath; that is, that there is no thing in-itself. As Bogdanov wrote,

> we take "physical nature" to be a *derivative* of complexes of an immediate nature (among which "psychical" coordinations are also included), and we take it to be a reflection of such immediate complexes in other coordinations that are analogous to those coordinations, only of a most complex type (i.e., coordinations that are the socially-organized experience of living beings).[21]

Consequently, truth does not pertain to the necessity to struggle to correlate the mind with mind-independent objects but, rather, in developing intersubjective consensus: "[t]he basis of 'objectivity' must lie in the sphere of *collective* experience."[22] Bogdanov also considered this to be consistent with contemporary studies of electricity, as he took this to mean that physics itself was increasingly proving that the fundament of matter was not the atom.[23]

As mentioned previously, while Lenin did write a series of articles against Bogdanov and the Russian Machists, his major statement in this polemic can be found in *Materialism and Empirio-Criticism*. The core of Lenin's position was to challenge the notion that Machism could be a philosophical foundation that was consistent with Marxist theory because of its defense of the concept of the identity of both consciousness and being. To the extent that capitalism is a global system, consciousness of capital and being of capital are, precisely, not identical. Furthermore, capitalism, like any class system, does not subsist on brute exploitation alone but on the enclosure of institutions around capital such that there appears to be no genuine alternative thereto. By assimilating consciousness with being, Machism risks simply being a way of affirming the contemporary order. As Lenin wrote,

> [s]ocial being and social consciousness are not identical, just as being in general and consciousness in general are not identical. From the fact that in their intercourse men act as conscious beings, it *does not follow* at all that social consciousness is identical with social being. In all social formations of any complexity—and in the capitalist social formation in particular—people in their intercourse *are not conscious* of what kind of social relations are being formed, in accordance with what laws they develop, etc. For instance, a peasant when he sells his grain enters into "intercourse" with the world producers of grain in the world market, but he is not conscious of it; nor is he conscious of the kind of social relations that are formed on the basis of exchange. Social consciousness *reflects* social being—that is Marx's teaching. A reflection may be an approximately true copy of the reflected, but to speak of identity is absurd. Consciousness in general *reflects* being—that is a general thesis of *all* materialism. It is impossible not to see its direct and *inseparable* connection with the thesis of historical materialism: social consciousness *reflects* social being.[24]

He maintained that it is necessary to posit a thing in-itself to recognize the gap between consciousness and being, which can only partially be overcome not through mere sensation, but action. To act means moving forward on the basis of incomplete knowledge, such that every situation entails risk and uncertainty. Moreover, the truth of the capitalist system does not come from its mere observation but, rather, through confrontation therewith; the truth of the inherent conflict within the system comes through instigating antagonism to see where the lines of power are truly drawn and from which something new could emerge.[25] For Lenin, the problems within the history and philosophy of science were animated by these very problems and dialectical materialism was the culmination of this trajectory.[26] It was this take on the scientific basis of Marxist theory that would win out in the October Revolution and that formed the basis of Soviet ideology about scientific socialism. Lenin also stressed that being a materialist in a dialectical sense means, on one hand, accepting the truth of mind-independent reality, while on the other hand, accepting that the conception of mind-independent reality will invariably change, as indicated previously.[27] Like the struggles of labor, the natural sciences are a human practice defined by the encounter between humanity and the limits of physical reality, meaning that its historical development is an essential part of the development of Marxism as scientific socialism.

By the 1920s, however, the newly founded Soviet Union found itself at a crossroads with respect to how this change in dialectical materialism was supposed to take place in the aftermath of the October Revolution and the long, bloody Russian Civil War. The Bolsheviks, now the Communist Party, had emerged as the nation's authority and following his death, Lenin's writings were now taken as the society's most sacred documents. However, Lenin's dialectical materialism did not provide unambiguous answers to all of the practical and theoretical problems that lay ahead, one such area being in the domain of science. Increasingly, developments in the natural sciences abroad began to present a challenge to the dialectical materialist approach to natural scientific theory. Quantum mechanics and relativity theory specifically appeared to pose a considerable challenge to the law of causality. Indeed, the two leading scientific minds of the day and respective representatives of quantum mechanics and relativity theory, Niels Bohr and Albert Einstein, would go on to confess that the bygone law of causality was done for. In his "The Quantum Postulate and the Recent Development of Atomic Theory," Bohr states that "[t]he quantum theory is characterized by the acknowledgement of a fundamental limitation in the classical physical ideas when applied to atomic phenomena. . . . The postulate implies a renunciation as regards the causal space-time co-ordination of atomic processes."[28] Likewise, in his "The Mechanics of Newton and Their Influence on the Development of Physics,"

Einstein asks "[w]ho would presume today to decide the question whether the law of causation and the differential law, these ultimate premises of the Newtonian view of nature, must definitely be abandoned?"[29] This placed the Soviet Union in a strange position: Either accept the new science and risk compromising the state ideology or accept the state ideology and risk falling behind in scientific pursuits. This debate culminated in the mechanist and Deborinite split of the late 1920s, with Hessen pursuing the latter position and arguing for the profound need to embrace both quantum mechanics and relativity theory.

QUANTUM MECHANICS

Among the two theories, we will turn our attention first to Hessen's approach to quantum mechanics. One can find Hessen's engagement with quantum mechanics throughout the entirety of his corpus but most explicitly in his Preface to the Russian translation of Austrian physicist, Arthur Haas's *Wave Mechanics and the New Quantum Theory*. Throughout his works, Hessen speaks to the many ways in which dialectical materialism already aligned with quantum mechanics, most importantly in the idea of the former that matter is simultaneously continuous and discontinuous and the idea in the latter that the electron behaves simultaneously as a particle and as a wave.[30] Where the two theories did not intersect, he argues, was due to certain shortcomings in quantum mechanics, as a result of its birth pangs that could be remedied by the adoption of a dialectical materialist approach. That approach involved seeing the continuation of particular recurring themes in the history of science in quantum mechanics: first, the law of causality and second, in individuals and collectives. In this section, I will show how Hessen attempted to lay the groundwork for a dialectical materialist conception of quantum mechanics along these two lines. To support my position, I will draw from Hessen's various scientific articles, paying special attention to his Preface to the Russian translation of *Wave Mechanics and the New Quantum Theory*.

The Law of Causality: Dynamical and Statistical

One of the central themes surrounding quantum mechanics early in the twentieth century was that of its apparent challenge to the central principle of any and all scientific research (i.e., the law of causality).[31] According to certain interpretations of quantum mechanics at the time, it appeared as though electrons were capable of moving entirely at random; English physicist, Paul Dirac even once wrote that nature "makes a free choice" in a paper from the Fifth Solvay Conference.[32] This observation conveyed the idea of what German physicist, Werner Heisenberg, would refer to as the "uncertainty principle," which Hessen interprets as the idea that "[a]n increased precision in the determination of a particle's position is related to an increased lack of precision in the determination of the momentum and vice versa."[33] From this principle, many physicists—from Bohr, Dirac, Eddington, Haas, and Lorentz to Schrödinger—entertained the notion that that the law of causality itself appeared to be at stake.[34] Take, for instance, Schrödinger's inaugural address upon his election to the Prussian Academy of Sciences, in which he stated that

> [t]he most burning issue for us today is whether we should abandon, along with classical mechanics, also its basis and method i.e. that immutable laws unambiguously determine the outcome in each individual case depending on arbitrary values of the initial conditions. It is a question of whether it is expedient to preserve the inviolability of the causality postulate.[35]

From this ambiguity, many physicists began to abandon any claims of establishing causal relationships in favor of a more parsimonious, descriptive approach. Others, however, saw the dawn of quantum mechanics as ushering in the age of a more idealistic or spiritualistic approach to physics more generally.[36] This can be observed in, for instance, the founding of the International Meta-Psychic Institute in Paris, France, the entertaining of the idea of founding of a Natural and Supernatural Society in England, as well as the fact that some of the most respected journals devoted to natural scientific research, such as *Nature*, began to publish a number of articles devoted to spiritualism.[37] Hessen notes how, within the Soviet Union, leaders within the Russian Orthodox Church were even undertaking to utilize theories of electromagnetism to explain the presence of Christ in the blood and wine of the Eucharist.[38]

For Hessen, however, this tendency toward idealism and spiritualism made perfect sense when understood as a process of emerging paradigms of materialism from a dialectical materialist standpoint. In this specific case, he argued that quantum mechanics was the sign that new ground was breaking in the understanding of two, long-competing conceptions of the law of causality: dynamical and statistical. A dynamical law, Hessen writes, "is primarily based on the complete and unambiguous determination of a subsequent state by the previous one,"[39] one example of which would be planetary orbit: "[a] planet's position is unambiguously and precisely determined by its previous position and speed."[40] Conversely, a statistical law, he notes, "is the supposition of no dependence of the subsequent state on the previous one;"[41] take, for instance, a coin toss: "[w]hen a coin is tossed for the tenth time, heads or tails show *totally irrespective* of the result of the ninth toss."[42]

For centuries, dynamical law always took precedent among physicists by virtue of the fact that such a law was seen as establishing a "necessary connection," whereas a statistical law was always seen as the placeholder for the hitherto undiscovered necessary connection.[43] This position is perhaps epitomized in what is known as "Laplace's demon":

> an intelligence which could comprehend all the forces by which nature is animated and the respective situation of the beings who compose it—an intelligence sufficiently vast to submit these data to analysis— . . . would embrace in the same formula the movements of the greatest bodies of the universe and those of the lightest atom; for it, nothing would be uncertain and the future, as the past, would be present to its eyes.[44]

Hessen points out, however, that throughout the course of the modern period, new discoveries made it increasingly unclear that dynamical law alone was constitutive of the law of causality. This came to light in the mid-nineteenth century with the ascent of thermodynamics.

The laws of thermodynamics were first explicated in their mature form by German physicist and mathematician, Rudolf Clausius, in 1850.[45] The 1st law, Clausius states, is that "[*t*]*he energy of the universe is constant*," while the second law is that "[*t*]*he entropy of the universe tends to a maximum*."[46] Taken together, they indicate that while the amount of energy in the universe remains constant, energy tends toward maximum entropy. These insights also led Clausius to an interesting conclusion about the origin and the destiny of the universe; namely, that if the first and second laws of thermodynamics were correct, then the universe would have had to have begun with an initial impulse and, conversely, would inevitably terminate in "heat death"—a position which has come to be known as "Clausius' paradox."[47]

Hessen maintained that there appeared to be an unresolved tension between so-called "reversible" and "irreversible" processes at the heart of Clausius' paradox.[48] One example of a reversible process would be that of a pendulum swing, where "[i]f we ignore the friction between the pendulum and the air and the thread upon which it is suspended, these oscillations, which correlate with the transformation from potential to kinetic energy, will occur indefinitely long, in one direction and then the other."[49] Conversely, an example of an irreversible process would be that of throwing a stone: "I throw a stone from a certain height. A stone raised to a certain height has a certain amount of (potential) energy. Upon falling to the ground, the stone produces a certain amount of heat, i.e. thermal energy, it makes a sound, i.e. vibrations in the air, a certain amount of mechanical energy, i.e. a visible deformation on the soil, etc."[50] Hessen described the conflict between reversible and irreversible processes in thermodynamics as follows:

> [if t]he thermal motion of molecules is a mechanical process[, and] . . . . all mechanical movements are *reversible*[,] . . . . [*w*]*hy, then, do thermal processes, which are nothing other than the mechanical movements of billions of molecules from the point of view of the kinetic theory of matter, behave as irreversible processes?*[51]

In other words, if molecular motion could be described as an example of a reversible process, why did it result in the irreversible process of increasing entropy? How could a reversible process serve as the basis of an irreversible one?

A resolution to Clausius' paradox arose in the studies of gases by figures like Scottish physicist, James Clerk Maxwell, and Austrian philosopher and physicist, Ludwig Boltzmann, both of whom share credit for introducing the so-called "statistical approach to molecular processes."[52] Boltzmann presented this statistical approach as a solution to the conflict of reversible and irreversible processes as follows:

> [w]e assume that molecules with different velocities are evenly distributed in space. In other words, if we set a certain velocity for a molecule, then molecules with that given velocity can be found in any point in space with the same probability. Molecules possessing a given velocity do not accumulate in one place, but are evenly distributed throughout the entire volume of the gas. . . . In a gas as a whole, in the aggregate of molecules, new laws appear that are specifically different from the purely mechanical laws that govern the actions of a single molecule. These laws are not the result of the equations of mechanics alone.[53]

The temperature of a gas, then, cannot be based on a single configuration of molecular particles. Indeed, temperature depends upon molecular particles being in motion and perpetually colliding with one another such that their position at any given moment is entirely random. That being the case, the relationship between the molecular level and the phenomenal level is not one of a necessary connection but of a combination of chance and necessity. Understood individually, molecules behave according to dynamical laws of motion, but understood together, the molecules behave according to statistical probability. For Boltzmann, taken together, this effectively resolved Clausius' paradox as it indicated that the relationship between the molecular level and the phenomenal level was not necessarily connected but was, instead, subject to a complex interplay of dynamical and statistical laws. This led to astonishing speculations that while it was more likely for an excitation of molecular particles to result in the release of heat, it was not wholly impossible for such an event to lead to freezing. This also provides a compelling objection to Clausius' account of the origins and ultimate fate of the universe.[54] The consequence of Boltzmann's approach was to suggest that dynamical laws alone did not express the law of causality, but that they were coequal with statistical laws.

Hessen points out that given the trajectory of the history of science, physicists should not have been all that surprised when quantum mechanics offered yet another turn of the screw:

> the peculiarity of the modern problem is that unlike in classical physics, statistical laws are not subordinate to the dynamical ones, and do not stand alongside them but present the only method that is currently available in physics and that is able to express the internal atomic phenomena's laws.[55]

For him, though, dialectical materialist philosophy presaged an answer to this problem via Engels's account of the unity in opposition of "chance" and "necessity." This is exemplified in a quote from Engels's work in which he writes, "[i]f the fact that a particular pea-pod contains six peas, and not five or seven, is of the same order as the law of motion of the solar system, or the law of the transformation of energy, then as a matter of fact chance is not elevated into necessity, but rather necessity degraded into chance."[56] In other words, if one tries to explain chance entirely via necessity or vice versa, one not only eliminates the opposite category, but one transforms the one category into the other. In this way, chance and necessity depend upon one another to be what they are in the first place. The same goes, Hessen contends, for dynamical and statistical laws. Any phenomenon depends upon the unity in opposition of dynamical and statistical laws, depending on what standpoint one adopts. They are not to the exclusion of one another but, on the contrary, are actually necessarily interdependent such that they present a much richer conception of the causal relationships between phenomena when understood together. Hessen asserts that "[t]herefore, if we reject the fatalistic concept of determinism on the one hand and accept chance as not simply a consequence of our ignorance but an objective category, then the opposition between dynamical and statistical laws is destroyed. They do not exclude but imply each other. They are both necessary and valid."[57] Both dynamical and statistical laws are abstractions; there are no circumstances in which simply one is at work at the expense of the other.[58] If we return, for instance, to the example of planetary orbits, then we find that they are in fact not such a simple expression of dynamical law. As Hessen writes that '[a] planet's motion is a dynamical law because we neglect its interaction with its environment. We would observe random variations similar to the ones we observe for a tossed coin if we look at an actual trajectory of a real planet and not at a trajectory of a physical point in mechanics; the aggregate of these variations can be expressed by a statistical law."[59] Likewise, concerning the example of coin tosses, we find that they too are not such a simple expression of statistical law. Hessen indicates that "the initial orientation, initial impulse, ejector mechanism, air movement as the coin falls and the whole complex set of initial conditions of a single toss[; e]ach single toss is fully determined by this set of conditions."[60] Thus, there is no such thing as a simple expression of either dynamical or statistical laws; all phenomena are intermixed.

Thus, when it comes to quantum mechanics, "[a] single quantum's direction of motion is accidental not in the sense that it is not determined but because a single quantum's behavior is not essential for the entire aggregate of quanta; only the whole identifies a statistical law."[61] That is, while a single quantum may appear to behave in a random fashion, it is nevertheless simply a member of a larger aggregate of quantum behavior that expresses a statistical law. This, of course, does not resolve every single conundrum of quantum mechanics, insofar as the issue of observability remains a problem because the scale of the particles and the instruments available to measure them mean that it is near impossible to determine any dynamical conception of quanta.[62] Nevertheless, Hessen argues that we should not treat this as an a priori problem of observation as such but as a limitation of the instruments of observation currently available and of current theories.[63]

Individuals and Collectives

Salvaging the law of causality meant not only reimagining the relationship between dynamical and statistical laws for Hessen but also that between individual and collective phenomena. The uncertainty principle, he argues, precipitated such an upheaval to physics in part because many physicists had remained wedded to the assumption that collective phenomena were nothing more than the sum of constituent individuals; that is, that wholes were nothing more than the sum of their parts. To contend this assertion, Hessen drew from the work of, among others, Richard von Mises. Von Mises himself was skeptical about the idea that quantum mechanics had ushered in the death knell of the law of causality, writing in *Probability, Statistics and Truth* that "[s]ome physicists feel that the ground has been cut from under their feet since the Uncertainty Principle was first announced. If no exact measurements are possible, not even in principle, what is the meaning of exact physical theories? In my opinion, these apprehensions are not justified."[64] He recognized that there are, in fact, two different types of properties that can be affiliated with aggregate phenomena: "additive" and "non-additive." Additive properties refer to those where "an aggregate as a whole is a simple sum of parts, and the laws for the whole are a sum of the laws for the parts"; one example of such a property would be something like "weight." Nonadditive properties, conversely, refer to those that "are not virtually contained in its individual components. They are manifested only in the whole and *qualitatively* differ from the properties of individuals;" one such example would be the properties of a gas which cannot be found in the constituent atoms.[65] Von Mises identifies those aggregates marked by nonadditive properties with a special term, namely, "collectives," or "a sequence of uniform events or processes which differ by certain observable attributes."[66] A collective is not simply the sum of constituent individuals but, rather, an aggregation of individuals that yields properties which arise only from the interrelationship between those components. In quantum mechanics, then, von Mises's insight suggests that if electron particles appear to behave randomly when observed as individuals, as members of a collective, their behavior can be discerned according to a concrete probabilistic value.[67] As he writes, "[t]he Uncertainty Principle derived in quantum mechanics implies that measurements in microphysics, like those in macrophysics, are elements of a collective; in either case, a vanishing variance of a measurement is merely the consequence of the choice of a sufficiently large unit of measurement."[68]

Hessen saw the pathway toward treating statistical law as not merely as a subjective phenomenon but as an objective phenomenon in von Mises's concept of the collective.[69] In this way, there were major parallels between von Mises's position and certain aspects of dialectical materialism:

> physics roughly follows the example and the direction of Marx, as we find in the "Introduction" to Marx's political economy, when he poses the question of the relationship between society and the individual, and the intention to build a theory of society by studying the individual in isolation. Marx calls this the age of large and small Robinsonades. And, of course, classical physics in its foundations was the physics of the large Robinsonades. But it limits the concept of a particular individual, forgetting further proof that an individual finds its full concept only in the aggregate, and that the problem of the individual and the aggregate can only be solved in dialectical materialism.[70]

Marx's precise critique of proponents of classical economics such as Adam Smith, David Ricardo, and so on, was that they grounded their studies in the assumption that human beings are atomized individuals who then come together to compose societies. For Marx, this was entirely backward, as historical analysis had revealed that the atomized individual was nowhere to be found at the beginning of human history but was, instead, a contemporary development in the historical process, specifically one invented around the time of industrial capitalism. For Marx, this phenomenon is epitomized in Daniel Defoe's 1719 novel, *Robinson Crusoe*. Hessen's own reference to "Robinsonades" comes from Marx's *Grundrisse*, a text which had yet to be published but which had nevertheless already been in circulation among

Soviet scholars in the early 1930s.[71] Hessen also finds this theme in a work by Plekhanov, titled *The Role of the Individual in History*.[72] In this text, Plekhanov responds to a common criticism of materialism, namely that it occludes any possibility of individual's possessing free will, by writing that these criticisms constructed a false dichotomy between voluntarism and determinism. An individual, he maintained, is not free to the extent that he/she acts wholly out of step with his/her time. Not even so-called "great individuals" can act outside of the possibilities available to them; he writes, "even at the height of his power Bismarck could not cause Germany to revert to natural economy."[73] Rather, individuals are free to the extent that they are able to shape the conditions available to them in their own time.

RELATIVITY THEORY

We will now turn our attention to Hessen's attempt to provide a dialectical materialist reading of relativity theory. Perhaps Hessen's best-known defenses of relativity theory can be found in his *The Main Ideas of the Theory of Relativity* along with his contributions to the *Soviet Encyclopedia.* Note that the first presentation of relativity theory can be traced back to Einstein's *annus mirabilis* paper, "On the Electrodynamics of Moving Bodies," in which he posits so-called "special relativity theory."[74] Special relativity is based on the two premises that the speed of light is constant and that the laws of nature are everywhere one and the same.[75] If these two premises are correct, he claims, one of the consequences would be that when observing an object emanating light, the speed of light emitted from that object would remain constant whether that object is still, moving toward, or moving away from the observer. This would only be possible, he argues, if space and time are capable of contracting and dilating; that is, if they were "relative."[76] In Hessen's day, some Soviet Marxists objected to this theory on the grounds that it appeared to mark a slippery slope toward philosophical relativism, a concern that they felt was only exacerbated by Einstein's flirtation with Machist epistemology.[77] Hessen maintains, however, that Einstein's philosophical position is not as clear cut as the aforementioned quotation might suggest, given that he effectively wavers between Machist epistemological and dialectical materialist tendencies in his work.[78] In many respects, he argues, relativity theory could certainly, and definitively, be severed from its Machist tendencies to be made compatible with dialectical materialism. Hessen sees the precedent for relativity theory in the history of science to the extent that he sees it as a continuation of competing conceptions of first, absolute and relative knowledge; second, space and time; third, mass and energy;

and fourth, ether. In this section, then, I will show how Hessen explains this compatibility along these four lines. To support my position, I will draw primarily from Hessen's *The Main Ideas of the Theory of Relativity* as well as his contributions to the *Soviet Encyclopedia.*

Absolute and Relative Knowledge

First and foremost, Hessen argues that relativity theory could be made consistent with dialectical materialism to the extent that they share an epistemological foundation, specifically in their understandings of unity in the opposition of absolute and relative knowledge. As indicated previously, certain Soviet Marxists objected to relativity theory because they believed it marked a slippery slope toward philosophical relativism.[79] For Hessen, however, this objection conflates two distinct conceptions of the notion of relativity (i.e., "philosophical" and "physical"), where the former refers to the relativity of knowledge and the latter refers to relativity as part of the objective structure of the fabric of reality.[80] In other words, whereas philosophical relativism treats knowledge as not extending beyond description, from the observer's standpoint, physical relativism, claims that given the nature of objects in space-time, knowledge of an object cannot be understood independently from its reciprocal relationship with the subject.[81] While the former is undoubtedly incompatible with dialectical materialism, the latter is in fact exemplary thereof.

Physical relativism, Hessen claims, is only incompatible with materialism to the extent that materialism is understood from a "metaphysical" or "mechanistic" standpoint, rather than from a dialectical one. According to the former, one can only cognize the object insofar as it is taken to be independent of the subject.[82] From a dialectical materialistic standpoint, however, one cannot omit the subject simply because the subject is itself material in nature. Hessen writes that "[t]he thinking process as the highest form of motion and the most complex one is a process in a human brain that is inseparably tied with other processes there and is another side of a *material* process."[83] In this way, the subject does not merely passively observe the object but affects the object through the very act of observation. Consequently, grasping the nature of the object does not mean extricating the subject but, rather, acquiring knowledge of the object through the subject. As he writes, "[a]ccording to dialectical materialism, a subject is both a condition and the only *method* for the consistent coverage and cognition of an object. The way towards absolute knowledge is through a subject."[84] He continues by claiming that

> [c]ognition is a process of reciprocal relationship between subject and object, these two opposites. Therefore, a subject and an object acquire a true reality, a

> living existence only in the *process of reciprocal relationship*. An object does not oppose a subject as some indifferent thing-in-itself. Not at all. It reveals itself to a subject through *reciprocal relationship*. It is cognisable, and this cognition is an endless process of reciprocal relationship. Engels says, 'We cannot go beyond cognition of this reciprocal relationship as there is nothing cognisable beyond it.'[85]

Contrary to objectors' claims, relativity theory and dialectical materialism share the idea that absolute knowledge is only possible from a relative standpoint. A reflection upon the nature of the subject, then, is not the obstacle to but actually asymptotically leads to absolute knowledge.[86]

The Space-Time Continuum

Second, Hessen argues that relativity theory could be made compatible with dialectical materialism, insofar as they share similar notions of the interrelatedness of space, time, matter, and motion. In relativity theory, space and time are neither independent of each other another nor are they independent of matter and motion, given that one of the central features of the theory is the claim that changes in the motion of matter warp space and either contract or dilate time.[87] For Hessen, dialectical materialism arrives at a similar point of view via the historical development of different philosophical conceptions of space and time.[88]

To defend his claim, Hessen advances the following argument. From the seventeenth to nineteenth centuries, the concepts of space and time could be understood from two opposing perspectives: the metaphysical or mechanistic materialist standpoint, as exemplified by English natural philosopher Sir Isaac Newton, and the so-called "idealistic" conception, as exemplified by German philosopher Immanuel Kant. Newton's natural philosophy, of course, is most fully elaborated in his magnum opus, the *Mathematical Principles of Natural Philosophy*, and it is in this text that Newton introduces the concept of "absolute space" along with "absolute time." "Absolute space," he explains is, "of its own nature without reference to anything external, always remains homogenous and immovable."[89] In other words, for Newton, if it were possible to remove all matter and motion from existence, space would remain as an absolute frame of reference, much like what Hessen calls an "infinitely large box without walls."[90] Without absolute space, motion would only ever be relative, such that there would be no basis upon which to distinguish motion from rest.[91] Newton makes a parallel argument with respect to time.[92] His natural philosophy can be identified as the metaphysical or mechanistic materialism that came to predominate classical physics because, while it recognizes the objective reality of matter, it nevertheless reduces all matter and motion to timeless mechanistic conceptions that fail to recognize the developmental aspects of both nature and concepts.[93]

Conversely, Kant expounds his critical philosophy most thoroughly in his own magnum opus, the *Critique of Pure Reason*, wherein he introduces the notion that both space and time are a priori forms of intuition. He states that "[w]e can accordingly speak of space, extended beings, and so on, only from the human standpoint. If we depart from the subjective condition under which alone we can acquire outer intuition, namely that through we may be affected by objects, then the representation of space signifies nothing at all."[94] That is, while Kant agrees with Newton that space and time are absolute frames of reference, they are not mind-independent realities but are, instead, projections of human consciousness onto the raw material of the world.[95] Kant's argument, effectively, is that we never encounter space and time through objects of sensation but that they are undeniably real as evidenced by our capacity to posit things like pure arithmetic or geometrical objects.[96]

The dialectical materialist conception of space and time, Hessen argues, effectively sublimates these two tendencies into one, first in the work of G. W. F. Hegel at the beginning of the nineteenth century and then in the work of Friedrich Engels at the century's close.[97] Hegel first argues for this position in the *Zusätse* of his *Philosophy of Nature*, in which he states that "[i]ts essence [motion] is to be the immediate unity of Space and Time; . . . Space and Time pertain to motion; the velocity, the quantum of Motion, is Space in relation to a specific Time elapsed,"[98] and likewise, that "Space and Time are filled with Matter. . . . Just as there is no Motion without Matter, so too, there is no Matter without Motion."[99] Engels, expanding upon Hegel's ideas, states that "[w]e know what an hour is, or a metre, but not what time and space are[.] As if time was anything other than just hours, and space anything but just cub. metres[.] The two forms of existence of matter are naturally nothing without matter, empty concepts, abstractions which exist only in our minds."[100] Not only can space not be understood independent of the interlocking constellation of matter, motion and time, but these notions are also the product of historical development.[101]

Space is an objective reality according to this conception, as Newton maintains, but it is also an expression of the relationship between the subject and the object as Kant maintains. Hessen writes as follows: "[m]otion is a synthesis of S. and time, because motion can only take place in time and S. Recognizing motion as the fundamental and inseparable property of matter, dialectic. materialism establishes not only the inseparability of space from time, but also the inseparability of time, space and matter."[102] While Hegel's and Engels' conception of space-time bore no direct influence on the relativistic conception of space-time,[103] Hessen maintains that it nevertheless provides grounds for showing that relativity theory is, in principle, compatible with dialectical materialism in a profoundly important sense.[104]

Mass-Energy Equivalence

Third, Hessen claims that relativity theory and dialectical materialism are compatible with respect to the concept of "mass-energy equivalence." While the notion of energy can be traced back to Aristotle's *energeia*, its modern conception is rooted in the work of seventeenth- and eighteenth-century German philosopher, G. W. Leibniz and his notion of *vis viva* [vital force].[105] It was not until the work of eighteenth- and nineteenth-century British scientist, Thomas Young, that the actual term "energy" came into use as "the ability to perform work due to speed acquired."[106] Lazare Carnot then posited the distinction between "living" and "latent living force," effectively introducing the concepts of "kinetic" and "potential" energy; these were terms that would later be popularized by Scottish mechanical engineer, William John Macquorn Rankine.[107] Among these terminological shifts, however, the meaning of the concept of energy remained fundamentally the same as "the ability to accomplish work,"[108] and, more specifically as "the measure of motion during its transformation from one form into another."[109] In another of his *annus mirabilis* papers, "Does the Inertia of a Body Depend Upon Its Energy Content?", Einstein introduces the notion that mass and energy are, in fact, equivalent, as he states that "[t]he mass of a body is a measure of its energy content."[110] Notably, some interpreters of relativity theory took this to mean that all matter is in fact reducible to energy, thereby making it the building block of physical reality; this is a position that can be found in the work of figures like James Rice, Arthur Eddington, and others.[111] For Hessen, however, mass-energy equivalence did not mean the reduction of matter to energy but, rather, recognizing that just as matter is inseparable from motion, so too is it inseparable from energy. Energy is not the fabric of reality but is, instead, a part of the unity in opposition with matter and motion.[112]

Hessen's argument proceeds as follows. By late in the nineteenth and early in the twentieth centuries, it was increasingly common for physicists to entertain the possibility that energy was the fabric of physical reality, a tendency exemplified in the development of a subfield of physics known as "energetics." Among the preeminent spokespeople of energetics were German mathematician and physicist, Georg Helm, and Latvian chemist and philosopher, Wilhelm Ostwald. In his *The Doctrine of Energy*, Helm encapsulates the basis of energetics with his claim that "[e]nergy is the true element of the world, for everything that we know of the world we know about energy."[113] Likewise, in the Preface to his *Lectures on Natural Philosophy*,

Ostwald summarizes it as "the way of uniting the concepts matter and mind by subordinating both to the concept of energy."[114] Energetics, however, was met with considerable opposition by Soviet philosophers and scientists alike, who saw it as favoring Machist epistemology; this was a position that they primarily arrived at via Lenin's *Materialism and Empirio-Criticism*. In this work, Lenin maintains that this slippery slope of energetics into idealism is in fact unsurprising. By regarding matter and motion as epiphenomenal of the relationship between the subject and the object, it effectively extricates the need to describe phenomena in objective terms at all. If objective phenomena are otherwise form-less, shape-less forces, then they cease to be conceivable independent of the observer's perspective, thereby collapsing into subjective idealism.[115] This was particularly concerning for Soviet philosophers and scientists, given that within the ranks of the Bolsheviks, Alexander Bogdanov's abandonment of "natural historical" materialism for energetic precipitated his own embrace of Machist epistemology.[116]

Hessen agreed with this objection to energetics, but he claimed that if one wanted to take a materialist approach to the concept of energy, then the mechanistic materialist paradigm was clearly insufficient. This approach seeks to explain all physical phenomena by explanations reduced to inert matter and mechanical motion.[117] While it continued to provide compelling explanations for numerous physical phenomena, it could not account for an increasing number of anomalies in the physics of the day: the principle of superposition, that magnetic phenomena that required the introduction of "hidden masses," that thermal motion required the introduction of molecular chaos, mass defect, radiation mass, and so on.[118] That mechanistic materialism cannot accommodate these phenomena is, for Hessen, unsurprising as it arose out of a socioeconomic formation with a narrower set of practical problems. He writes that

> E. does not play any role in Newtonian mechanics. This is primarily due to the fact that all of Newton's mechanics mainly interprets the laws of purely mechanical motion, and does not pose the problem of transforming the mechanical form of E. into other forms. This interpretation of Newton can be explained by the fact that the socio-economic formation of his era posed purely mechanical problems as the main tasks of physics. As soon as the steam engine began to be developed after the industrial revolution, and with it thermodynamics, the problem of transforming the thermal form of motion into the mechanical one came to fruition (see *Entropy*). The development of this problem was stimulated by the gradual highlighting of the electrical and chemical forms of motion. This led to the final formulation of the "law of E. conservation."[119]

Newtonian mechanics, as an expression of mechanistic materialist physics, partly arose from the study of machines which were powered by human or natural sources (i.e., which only concerned the relationship between different forms of mechanical motion). The problem arose of transforming thermal energy into mechanical motion when the steam engine came to be incorporated into economic production, with the rise of industrial capitalism; problems were exacerbated further with the harnessing of electromagnetism, radiation, and so on.[120] Ultimately, these attest to the fact that the practical problems of the socioeconomic formation of the day had yet to be sufficiently grasped by theory.

In fact, Hessen notes that energy is a concept that exceeded the limits of mechanistic materialism from its very inception because it expressly did not arise within the problematic of mechanistic materialism at all. One cannot ignore the fact that the history of modern physics, he indicates, is the product of the unity in opposition between two distinct lines of thought, one running from Newton to Helmholtz and the other from Leibniz to Mayer.[121] According to the former, matter is inherently inert and can only be acted upon, while matter is inherently active according to the latter.[122] The former proceeds effectively unchanged from Newton to Helmholtz, Helmholtz writes that "[t]he existence of matter in itself is to us something tranquil and devoid of action,"[123] and, therefore, that "[t]he phænomena of nature are to be referred back to motions of material particles possessing unchangeable moving forces, which are dependent upon conditions of space alone."[124] Contrary to Newton, Leibniz maintains that matter is active and that it possesses a force of its own.[125] Leibniz discovers this through his critique of the Cartesian notion of the principle of the constancy of motion, writing that

> the *foundation of the laws of nature* . . . is not to be found in the principle that the same quantity of motion is conserved . . ., but rather in the principle that the *same quantity of active power is necessarily conserved*, and indeed—a thing which I discovered happens for a most beautiful reason—the *same quantity of motive action* also, the value of which is far different from what the Cartesians think of as the quantity of motion.[126]

The concept of energy, then, arose from Leibniz's philosophico-theological deduction of the concept of *vis viva*.[127] The materialist interpretation of energy was then confirmed a century later through Mayer's experimentation with living organisms, among other things.[128] Mayer maintains that matter is not reducible to mechanism but that all forms of motion can be converted one into the other; this was a principle that he identified as a universal law of nature. As he states, "[i]t can be proven *a priori* and confirmed everywhere by experience that the different forces can be *transformed* into one another;"[129] likewise, that "[t]o become acquainted with force in its different *forms*, to investigate the conditions of its *metamorphoses*—this is the only task of physics."[130]

For Hessen, establishing these competing lineages provides a foothold by which to get a better grasp on the concept of energy in the present; that is, that it is suspended within a larger debate about the continuity or discontinuity of physical reality. This problematic concerns the fact that energy expresses both the quantitative continuity between different forms of motion (i.e., mechanical, thermal, electromagnetic, etc.) *and* the qualitative discontinuity between them.[131] In early twentieth-century physics, the concept of energy was being pulled in two apparently irreconcilable directions. Relativity theory appeared to follow the Newton–Helmholtz line of favoring continuity, by aiming to prove that energy was inherently continuous, whereas quantum mechanics appeared to follow the Leibniz–Mayer line, by aiming to prove that energy was inherently discontinuous.[132] As he writes, however, historical analysis goes to show that perhaps the answer does not lie in just one of these answers:

> [t]he one-sided emphasis on the discontinuity of energetic processes and the mechanical transfer of the concept of the discontinuity of E. generally leads to an incorrect approach to the problem of E. and matter, since they overlook the main problem of the synthesis of continuity and discontinuity. The entire development of modern physics shows the limited point of view of reducing all physical phenomena to either pure discontinuity or pure continuity. On the contrary, the correct interpretation of the latest changes taking place in the theory of the structure of matter . . . confirms the point of view of dialectical materialism regarding the synthesis of opposites; that is, of discontinuity and continuity.[133]

In other words, energy is inherently defined by the unity in opposition between qualitative discontinuity and quantitative continuity between states of matter and motion.[134] Contrary to the energeticists, energy is not the principle of matter and motion, but as previously indicated, "the measure of motion during its transformation from one form to another."[135] Thus, he maintains that treating mass and energy as equivalent does not mean treating energy as the foundational principle of physics but, rather, as the actuality or potentiality of all matter and motion to be converted into other forms of matter and motion.[136]

Ether

Fourth, Hessen argues that relativity theory could be made to be consistent with dialectical materialism, insofar as both theories offer compatible notions of ether. By early in the twentieth century, the concept of ether had admittedly been deemed anachronistic by virtue of Einstein's positing of special relativity. Hessen's own willingness to entertain the concept was in fact met with considerable ridicule in the Soviet context, even among his allies who were favorable to relativity theory.[137] Nevertheless, he maintained that the concept of ether was consistent with general relativity theory, the fact that it required some revision notwithstanding, and might even hold the key to resolving the tension between quantum mechanics and relativity theory.[138]

The concept of ether dates back millennia, but in classical physics, it was understood as "a special continuous medium in which waves can propagate and which is a carrier of an electric and magnetic field."[139] This conception spanned from roughly the seventeenth to twentieth centuries. It came into more widespread use with the popularization of the "wave theory" of light, first elaborated in 1818 by French civil engineer and physicist, Augustin-Jean Fresnel, in his *Memoir on the Diffraction of Light*. Later in the nineteenth century, English physicist, Michael Faraday, and Scottish physicist, James Clerk Maxwell, denied the possibility of action at a distance for the propagation of electromagnetic waves, leading them to also posit an invisible medium through which the transmission of such waves was possible. Like sound waves through air, light, and electromagnetic waves needed a medium through which to pass. If ether existed, however, it had to be understood as fundamentally different from the energy and matter which it transmitted; namely, it could neither possess any mass, nor appear to even move and it did not appear to obey the law of gravity. Based upon the Lorentz theorem, physicists determined that ether could be understood as an absolute inertial frame of reference by which to measure motion and rest, and by the late nineteenth and early twentieth centuries, they began to conduct experiments to prove its existence and to determine its basic properties.[140]

One such effort to prove the existence of ether was that of measuring so-called "ether wind." As with objects traveling through air or water, the passage of objects through the ether was believed to lead to similar such distortions in the medium. American physicists Albert A. Michelson and Edward W. Morley carried out an experiment to measure the ether wind in which they projected and reflected beams of light at perpendicular angles, hoping that they could determine the overall interference of the ether by measuring the differences in the speed of light. However, the so-called Michelson-Morley experiment would, in fact, prove deleterious to the concept of ether altogether, given the reflected light rays traveled at identical speeds.[141] In their paper "On the Relative Motion of the Earth and the Luminiferous Ether," they summarized the results of their experiment by suggesting that if there was an ether wind, it was so small as to have escaped their detection: "[i]t appears . . . reasonably certain that if there be any relative motion between the earth and the luminiferous ether, it must be small."[142] Further experiments by Michelson and Morley and others by Dayton Miller, however, would suggest that this experiment eventually came to be seen as sounding the death knell to ether theory.[143] For Einstein, the experiment proved that ether was a "superfluous"[144] concept.[145]

Hessen would argue, however, that special relativity left open the question of how to understand the transmission of both light and electromagnetic waves through space, along with the question of the objectivity of both motion and rest.[146] This was why he argues that not only did some scientists like J. J. Thomson continue to entertain the classical conception of ether but also why Einstein himself would reintroduce the concept in his discussions of general relativity theory.[147] As Einstein remarks in a lecture from 1920, "Ether and The Theory of Relativity," "according to the general theory of relativity space is endowed with physical qualities; in this sense, therefore, there exists an ether. According to the general theory of relativity space without ether is unthinkable."[148] In other words, the jury was far from in concerning the question of ether.

Hessen maintains that a dialectical materialist standpoint could overcome the previously espoused and erroneous conceptions of ether which presupposed that matter must be inherently "heavy and inert" and, likewise, continuous. From this standpoint, the concept of matter would always be subject to change and, hence, to a broadening of understanding of the conception of matter, given that this would yield a better understanding of ether.[149] As Engels writes, "[i]f it *exists* at all, it must be of a material nature, it must come under the concept of matter."[150] Thus, Hessen maintains that

> [t]he primary methodological error of general relativity theory is that it considers E. to be an absolutely continuous medium, while all modern theories of the structure of matter (quantum theory, wave mechanics) more and more confirm the position of dialectical materialism, according to which matter in its objective structure is simultaneously continuous and discontinuous.[151]

With further elaboration, then, he felt that the notion of ether held the key to some of the central questions of modern physics.

NOTES

1. This chapter contains passages from my "Individuals and Collectives in the Philosophy of Boris Hessen: An Introduction," *Science in Context* 34.1 (2021): 121–36, as well as my "A Pantheon of Great Ideas: Boris Hessen and the History and Philosophy of Science," in *Manuscripts and Documents on the History of Physics:*

A Historical Materialist Textbook, ed. Pietro Omodeo and Sean Winkler (Venice: Verum Factum, 2022), 33–73.

2. Bochenski, *Soviet Russian Dialectical Materialism [Diamat]*, 34.

3. See second section of this chapter.

4. Hessen and Egorshin, "On Cde. Timiryazev's Attitude towards Contemporary Science," 146–147.

5. For example, Marx and Engels wrote in the *Communist Manifesto* that "Critical-Utopian Socialism and Communism . . . . dream of experimental realization of their social Utopias . . . and to realize all these castles in the air, they are compelled to appeal to the feelings and purses of the bourgeois. By degrees they sink into the category of the reactionary [or] conservative Socialists depicted above, differing from these only by more systematic pedantry, and by their fanatical and superstitious belief in the miraculous effects of their social science." Similarly in *Socialism: Utopian and Scientific*, Engels wrote that upon Marx's and his discovery, "[f]rom that time forward Socialism was no longer an accidental discovery of this or that ingenious brain, but the necessary outcome of the struggle between two historically developed classes—the proletariat and the bourgeoisie. Its task was no longer to manufacture a system of society as perfect as possible, but to examine the historic-economic succession of events from which these classes and their antagonism had of necessity sprung, and to discover in the economic conditions thus created the means of ending the conflict." K. Marx and F. Engels, *Manifesto of the Communist Party*, in *Marx and Engels: Collected Works, vol. 6; Marx and Engels 1845–48*, ed. R. Daglish, et al. and trans. J. Cohen, et al. (London: Lawrence and Wishart, 2010 [1848]), 516–17; F. Engels, *Socialism: Utopian and Scientific*, in *Marx and Engels: Collected Works, vol. 24; Marx and Engels 1874–83*, ed. Y. Kalinina and trans. D. Forgacs, et al. (London: Lawrence and Wishart, 2010 [1880]), 304–5.

6. L. R. Graham, *Science, Philosophy, and Human Behavior in the Soviet Union* (New York: Columbia University Press, 1987), 25.

7. J. Dietzgen, *Excursions of a Socialist in the Domain of Epistemology*, in *Some of the Philosophical Essays by Joseph Dietzgen*, trans. M. Beer and T. Rothstein (Chicago: Charles H. Kerr, 1906). See also Graham, *Science, Philosophy, and Human Behavior in the Soviet Union*, 25.

8. G. Plekhanov, *The Materialist Conception of History*, trans. unknown (New York: International Publishers, 1940 [1897]). See also Graham, *Science, Philosophy, and Human Behavior in the Soviet Union*, 25.

9. D. Bakhurst, "On Lenin's *Materialism and Empirio-Criticism*," *Studies in East European Thought* 70 (2018): 108.

10. Lenin, *Materialism and Empirio-Criticism*, 21; W. Hedeler, "Lenin and Stalin, Theory and Politics," in *The Palgrave Handbook of Leninist Philosophy*, ed. T. Rockmore and N. Levine (London: Palgrave Macmillan, 2018), 260.

11. E. Ilyenkov, *Leninist Dialectics and the Metaphysics of Positivism*, trans. unknown (Online, 1982), www.marxists.org.

12. Ibid.

13. V.I. Lenin, "A Letter to A.M. Gorky (25.02.1908)," in V. I. Lenin: Collected Works, vol. 13; June 1907–April 1908, ed. C. Dutt and trans. B. Isaacs (Moscow:

Progress Publishers, 1978 [1908]), 449, 532n156. See also D. Steila, "Lenin's Philosophy in Intellectual Context," in *The Palgrave Handbook of Leninist Philosophy* (London: Palgrave Macmillan, 2018), 97–99.

14. Ilyenkov, *Leninist Dialectics and the Metaphysics of Positivism.*

15. A. A. Богданов, et al., Очерки по философии марксизма (Москва: Звено, 1910).

16. Ilyenkov, *Leninist Dialectics and the Metaphysics of Positivism*, Online.

17. A. A. Богданов, «Развитие жизни в природе и обществе», in *Из психологии общества* (Санкт-Петербург: Электропечатня товарищества «Дело», 1904), 35. See also Lenin, *Materialism and Empirio-Criticism*, 322.

18. R. Avenarius, *Kritik der Reinen Erfahrung, Bände 1–2* (Leipzig: O.R. Reisland, 1921). See also A. Bogdanov, *Empiriomonism: Essays in Philosophy, Books 1–3*, ed. and trans. D. G. Rowley (Leiden: Brill, 2020 [1904–1906]), 6.

19. E. Mach, *Contributions to the Analysis of the Sensations*, trans. C. M. Williams (Chicago: The Open Court Publishing Company, 1897 [1886]), 10; R. Avenarius, *Prolegomena zu einer Kritik der reinen Erfahrung* (Leipzig: Fues, 1876), 65. See also Lenin, *Materialism and Empirio-Criticism*, 43, 49.

20. Bogdanov, *Empiriomonism*, 296–97. See also Lenin, *Materialism and Empirio-Criticism*, 57–58.

21. Bogdanov, *Empiriomonism*, 103. See also Lenin, *Materialism and Empirio-Criticism*, 228.

22. Bogdanov, *Empiriomonism*, 273, 18. See also Lenin, *Materialism and Empirio-Criticism*, 123, 124.

23. Bogdanov, *Empiriomonism*, 279. See also Lenin, *Materialism and Empirio-Criticism*, 272.

24. Ibid., 323.

25. V.I. Lenin, *The Economic Content of Narodism and the Criticism of It in Mr. Struve's Book (The Reflection of Marxism in Bourgeois Literature)*, in *V.I. Lenin: Collected Works, vol. 1; 1893–1894*, ed. and trans. Institute of Marxism-Leninism, Central Committee of the CPSU (Moscow: Progress Publishers, 1977 [1895]), 400. See also Steila, "Lenin's Philosophy in Intellectual Context," 90.

26. Lenin, *Materialism and Empirio-Criticism*, 130.

27. See chapter 2.

28. N. Bohr, "The Quantum Postulate and the Recent Development of Atomic Theory," *Nature* (1928): 580. See also Б. М. Гессен, «Идеалистические течения в современной физике и борьба с ними», *Молодая гвардия* 3 (1929): 64.

29. Einstein, "The Mechanics of Newton and Their Influence on the Development of Theoretical Physics," 261. See also Гессен, «Идеалистические течения в современной физике и борьба с ними», 64; Hessen, "Preface to Articles by A. Einstein and J.J. Thomson," 94.

30. B. Hessen, "On the Question of the Causality Problem in Quantum Mechanics. Preface to the 1931 Russian Translation of Arthur Haas, *Materiewellen und Quantenmechanik*," in *Boris Hessen: Physics and Philosophy in the Soviet Union, 1927–1931; Neglected Debates on Emergence and Reduction*, ed. and trans. C. Talbot and O. Pattison (Cham: Springer Nature, 2021 [1931]), 143.

31. Ibid., 144–51.

32. Ibid., 151. As mentioned in the editors'/translators' note, "free'" does not appear in the original. See also G. Bacciagaluppi and A. Valentini, *Quantum Theory at the Crossroads: Reconsidering the 1927 Solvay Conference* (Cambridge: Cambridge University Press, 2009), 405.

33. Hessen, "On the Question of the Causality Problem in Quantum Mechanics," 152.

34. Ibid., 144, 145, 151, 152, 153.

35. Ibid., 152–53. See also E. Schrödinger, "Sitzungsberichte der Preussischen Akademie der Wissenschaften, 1929," in *Gesammelte Abhandlungen, T. 3* (Wien: Verlag der Österreichischen Akademie der Wissenschaften, F. Vieweg, 1984 [1929]), 304.

36. Гессен, «Идеалистические течения в современной физике и борьба с ними», 64–67.

37. Ibid., 66.

38. Ibid., 70.

39. Hessen, "On the Question of the Causality Problem in Quantum Mechanics," 148.

40. Ibid.

41. Ibid.

42. Ibid.

43. Ibid., 149.

44. P.-S. Marquis de Laplace, *A Philosophical Essay on Probabilities*, trans. F.W. Truscott and F.L. Emory (New York: John Wiley and Sons, 1902 [1814]), 4. See also B. M. Hessen, "Mechanical Materialism and Modern Physics," trans. S. Winkler and rev. A. Kojevnikov, *Science in Context* 34.1 (2021): 169.

45. Hessen, "Mechanical Materialism and Modern Physics," 162.

46. R. Clausius, *The Mechanical Theory of Heat, with Its Applications to the Steam-Engine and to the Physical Properties of Bodies*, ed. and trans. T.A. Hirst (London: John van Voorst, 1867), 365.

47. Hessen, "Mechanical Materialism and Modern Physics," 166.

48. Ibid, 166–167.

49. Ibid, 165.

50. Ibid, 164.

51. Ibid, 167. See also B. M. Hessen, "Marian Smoluchowski (On the 10th Anniversary of His Death)," trans. S. Winkler and rev. A. Kojevnikov, *Science in Context* 34.1 (2021): 138.

52. Hessen, "Mechanical Materialism and Modern Physics," 167; Hessen, "Marian Smoluchowski," 138.

53. Hessen, "Mechanical Materialism and Modern Physics," 167.

54. Hessen, "Marian Smoluchowski," 138. See also Hessen, "Mechanical Materialism and Modern Physics," 183.

55. Hessen, "On the Question of the Causality Problem in Quantum Mechanics," 149.

56. F. Engels, *Dialectics of Nature*, in *Marx and Engels: Collected Works, vol. 25; Engels*, ed. N. Karmanova, et al. and trans. C. Dutt (London: Lawrence and Wishart, 2010 [1883]), 498–501. See also Hessen, "On the Question of the Causality Problem in Quantum Mechanics," 150–51; Hessen, "Preface to Articles by A. Einstein and J.J. Thomson," 97.

57. Hessen, "On the Question of the Causality Problem in Quantum Mechanics," 151.

58. Ibid., 150.

59. Ibid.

60. Ibid., 148.

61. Ibid., 149, 151.

62. Ibid., 154.

63. Ibid., 155.

64. R. von Mises, *Probability, Statistics and Truth*, trans. H. Geiringer (London: George Allen and Unwin Ltd., 1957 [1928]), 216–17.

65. Hessen, "Selected Material on the Work of Richard von Mises," 135.

66. von Mises, *Probability, Statistics and Truth*, 12.

67. Ibid.

68. Ibid., 223.

69. Ienna and Rispoli, "Boris Hessen at the Crossroads of Science and Ideology," 53; Pechenkin, *L. I. Mandelstam and His School in Physics*), 98; Корсаков, Козенко, and Грачева, Борис Михайлович Гессен, 63.

70. Б.М. Гессен, «Выступление на научной сессии философии, посвящённой 25-летию выхода в свет труда В.И. Ленина «*Материализм и эмпириокритицизм*». 22 Июня 1934 г.», ed. С.Н. Корсаков, *Эпистемология и философския науки* 56.1 (2019 [1934]): 223.

71. In the *Grundrisse*, Marx writes as follows: "[t]he individual and isolated hunter and fisherman, who serves Adam Smith and Ricardo as a starting point, is one of the unimaginative fantasies of the 18th century. Robinsonades which, contrary to the fancies of the history of civilization, by no means signify simply a reaction against over-refinement and a reversion to a misconceived natural life. No more is Rousseau's *contrat social*, which by means of a contract establishes a relationship and connection between subjects that are by nature independent, based on this kind of naturalism. This is an illusion and nothing but the aesthetic illusion of the small and big Robinsonades. It is, rather, the anticipation of 'bourgeois society', which began to evolve in the 16th century and was making giant strides towards maturity in the 18th. . . . The prophets of the 18th century . . . . saw this individual not as an historical result, but as the starting point of history; not as something evolving in the course of history, but posited by nature, because for them this individual was the natural individual, according to their idea of human nature. This delusion has been characteristic of every new epoch hitherto." K. Marx, *Economic Manuscripts of 1857–58*, in *Marx and Engels: Collected Works, vol. 28; Marx 1857–61*, ed. L. Miskievich and trans. E. Wangermann (London: Lawrence and Wishart, 2010 [1857–1858]), 17–18. Marx also mentions the story of Robinson Crusoe in *Capital, vol. 1.*, 87–88.

72. Hessen, "Preface to Articles by A. Einstein and J.J. Thomson," 93; Hessen, "Marian Smoluchowski," 140; Hessen, "Mechanical Materialism and Modern Physics," 179. In *The Role of the Individual in History*, Plekhanov writes as follows: "who makes history? It is made by the *social man*, who is its *sole 'factor'*. The social man creates his own, social, relationships. But if in a given period he creates given relationships and not others, there must be some cause for it, of course; it is determined by the state of his productive forces. No great man can foist on society relations which *no longer* conform to the state of these forces, or which *do not yet* conform to them. In this sense, indeed, he cannot make history, and in this sense he would advance the hands of his clock in vain; he would not hasten the passage of time, nor turn it back. . . . Social relationships have their inherent logic: as long as people live in given mutual relationships they will feel, think and act in a given way, and no other. Attempts on the part of public men to combat this logic would also be fruitless; the natural course of things (the logic of social relationships) would reduce all his effort to naught. But if I know in what direction social relations are changing owing to given changes in the social-economic process of production, I also know in what direction social mentality is changing; consequently, I am able to influence it. Influencing social mentality means influencing historical events. Hence, in a certain sense, I *can make history*, and there is no need for me to wait while 'it is being made.'" Plekhanov, *The Role of the Individual in History*, 60–61.

73. Plekhanov, *The Role of the Individual in History*, 60.

74. A. Einstein, "On the Electrodynamics of Moving Bodies," in *The Collected Papers of Albert Einstein, vol. 2: The Swiss Years – Writings, 1900–1909*, ed. A. Beck and P. Havas (Princeton: Princeton University Press, 1989 [1905]), 141.

75. Ibid., 140.

76. Ibid., 144–45, 153.

77. A. Pais, *Subtle is the Lord: The Science and the Life of Albert Einstein* (Oxford: Oxford University Press, 2005), 283.

78. H. M. Müntz and B. M. Hessen, "Einstein," translated by S. Winkler and revised by O. Bashkina, 155 (see Appendix 6).

79. B. Hessen, Selections from *The Main Ideas of the Theory of Relativity*, in *Boris Hessen: Physics and Philosophy in the Soviet Union, 1927–1931; Neglected Debates on Emergence and Reduction*, ed. and trans. C. Talbot and O. Pattison (Cham: Springer Nature, 2021 [1928]), 118.

80. Ibid., 122.

81. Ibid., 120.

82. Ibid., 119.

83. Ibid., 115–16.

84. Ibid., 119.

85. Ibid., 120.

86. Ibid., 119, 121.

87. B. M. Hessen, "Space," translated by S. Winkler and revised by O. Bashkina, 115 (see Appendix 1).

88. Ibid.

89. I. Newton, *The Principia: Mathematical Principles of Natural Philosophy*, trans. I. Bernard Cohen, Anne Whitman and Julia Budenz (Berkeley: University of California Press, 1999 [1687]), 408.

90. Hessen, "Space," 115.

91. Newton provides support for this position in the General Scholium of the *Principia*, which features his famous 'bucket' thought experiment. Newton, *The Principia*, 412–13.

92. Ibid., 410.

93. Hessen, "Space," 114.

94. I. Kant, *Critique of Pure Reason*, ed. and trans. P. Guyer and A.W. Wood (Cambridge: Cambridge University Press, 1998 [1781]), 159–60.

95. Hessen, "Space," 114–115.

96. I. Kant, *Prolegomena to Any Future Metaphysics*, 2nd ed., trans. J.W. Ellington (Indianapolis: Hackett Publishing Company, Inc., 2001 [1783]), 25.

97. Hessen, "Space," 114–115.

98. G. W. F. Hegel, *Philosophy of Nature; Encyclopaedia of the Philosophical Science, Pt. 2 (1830)*, trans. A.V. Miller (Oxford: Clarendon Press, 2004 [1830]), 43. I maintain that this is likely Hessen's source for treating Hegel as an antecedent to a dialectical materialist conception of space, because Hessen frequently cites the "Notes and Fragments" of Engels' *Dialectics of Nature*, which includes a citation of Hegel's *Philosophy of Nature*. Hessen himself went on to review the Russian translation of Hegel's *Philosophy of Nature* in 1935. See also Engels, *Dialectics of Nature*, 524; Б.М. Гессен, Рецензия на Г.В.Ф. Гегель Сочинения, Т. *2: Философия природы, За большевистскую книгу* 13/14 (1935): 25–29.

99. Hegel, *Philosophy of Nature*, 43. See also Engels, *Dialectics of Nature*, 524.

100. Engels, *Dialectics of Nature*, 515. See also Hessen, "Space," 114.

101. Hessen, "Space," 114.

102. Ibid.

103. Ibid.

104. Ibid.

105. B. M. Hessen, "Energy," translated by S. Winkler and revised by O. Bashkina, 122 (see Appendix 3).

106. Note that while Hessen attributes this quote to Young, Young himself actually defines energy as "the tendency of a body to ascend or to penetrate to a certain distance, in opposition to a retarding force." T. Young, *A Course of Lectures on Natural Philosophy and the Mechanical Arts, vol. 1* (London: William Savage, 1807), 44.

107. Hessen, "Energy," 122.

108. Ibid, 120.

109. Ibid. See also Engels, *Dialectics of Nature*, 477.

110. A. Einstein, "Does the Inertia of a Body Depend Upon Its Energy Content?" in *The Collected Papers of Albert Einstein, vol. 2: The Swiss Years – Writings, 1900–1909*, ed. A. Beck and P. Havas (Princeton: Princeton University Press, 1989 [1905]), 174.

111. Гессен, «Идеалистические течения в современной физике и борьба с ними», 70; Hessen, "Entropy," translated by S. Winkler and revised by O. Bashkina, 149 (see Appendix 4).

112. Hessen, "Energy," 135.

113. G. Helm, *Die Lehre von der Energie – Historich-Kritisch Entwickelt: Nebst Beiträgen zu einer allgemeinen Energetik* (Leipzig: Verlag von Arthur Felix, 1887), 56–57. See also R. J. Deletete, "*Die Lehre von der Energie*: Georg Helm's Energetic Manifesto," *Centaurus* 47 (2005): 140–62

114. W. Ostwald, *Vorlesungen über Naturphilosophie, vol. 2* (Leipzig: Verlag von Veit and Comp., 1902), viii. See also Lenin, *Materialism and Empirio-Criticism*, 271.

115. Lenin, *Materialism and Empirio-Criticism*, 271.

116. Ibid., 231.

117. Hessen, "Energy," 123–124.

118. Ibid, 126–127.

119. Ibid, 130–131.

120. Ibid, 131.

121. Ibid, 129–133.

122. Ibid.

123. H. von Helmholtz, "On the Conservation of Force; a Physical Memoir [Read before the Physical Society of Berlin on the 23rd of July, 1847. Berlin. G. Reimer]," in *Scientific Memoirs, Selected from The Transactions of Foreign Academies of Science and from Foreign Journals, vol. 1*, ed. J. Tyndall and W. Francis and trans. J. Tyndall (London: Taylor and Francis, 1853 [1847]), 115. See also Hessen, "Energy."

124. von Helmholtz, "On the Conservation of Force," 116. See also Engels, *Dialectics of Nature*, 370n; Hessen, "Energy," 131.

125. Hessen draws this from a passage in Lenin's "Conspectus of Feuerbach's Book *Exposition, Analysis and Critique of the Philosophy of Leibnitz*," in the *Philosophical Notebooks*, where Feuerbach writes that "'for Leibnitz, corporeal substance is no longer, as for Descartes, a merely extended dead mass, brought into motion from outside, but as substance it has within it an active force, a never-resting principle of activity.'" Lenin, "Conspectus of Feuerbach's Book *Exposition, Analysis and Critique of the Philosophy of Leibnitz*," 378. See also Hessen, "Energy," 131.

126. G.W. Leibniz, "On Nature Itself, or on the Inherent Force and Actions of Created Things, 1698," in *Philosophical Papers and Letters*, 2nd ed., ed. and trans. L. E. Loemker (Dordrecht: Kluwer Academic Publishers, 1989 [1698]), 499. See also Hessen, "Energy," 130.

127. Hessen also draws this from a passage found in Lenin's "Conspectus of Feuerbach's Book *Exposition, Analysis and Critique of the Philosophy of Leibnitz*" in the *Philosophical Notebooks*, where Lenin writes that "Leibnitz through theology arrived at the principle of the inseparable (and universal, absolute) connection of matter and motion." Lenin, "Conspectus of Feuerbach's Book *Exposition, Analysis and Critique of the Philosophy of Leibnitz*," 377. See also Hessen, "Energy," 130.

128. Lenin, "Conspectus of Feuerbach's Book *Exposition, Analysis and Critique of the Philosophy of Leibnitz*," 377. See also Hessen, "Energy," 130.

129. J. R. von Mayer, *Die organische Bewegung in ihrem Zusammenhange mit dem Stoffwechsel. Ein Beitrag zur Naturkunde* (Heilbronn: C. Drechsler, 1845), 6. See also K. L. Caneva, *Robert Mayer and the Conservation of Energy* (Princeton: Princeton University Press, 1993), 60; Hessen, "Energy," 131.

130. von Mayer, *Die organische Bewegung in ihrem Zusammenhange mit dem Stoffwechsel*, 6. See also Caneva, *Robert Mayer and the Conservation of Energy*, 198; Hessen, "Energy," 131.

131. Hessen, "Energy," 131–132.

132. Ibid.

133. Ibid, 129.

134. Ibid.

135. Ibid, 120. See also Engels, *Dialectics of Nature*, 477.

136. Hessen, "Energy," 135.

137. See chapter 1.

138. B. M. Hessen, "Ether," translated by S. Winkler and revised by O. Bashkina, 118 (see Appendix 4).

139. Ibid, 117.

140. Ibid, 117–118.

141. E. T. Whittaker, *A History of the Theories of Aether and Electricity, vol. 1: From the Age of Descartes to the Close of the Nineteenth Century* (London: Longmans, Green and Co., 1910), 115.

142. A. A. Michelson and E. W. Morley, "On the Relative Motion of the Earth and the Luminiferous Ether," *The American Journal of Science* 34.203 (1887): 341.

143. Hessen, "Ether," 117–118.

144. Einstein, "On the Electrodynamics of Moving Bodies," 141.

145. E.T. Whittaker, *A History of the Theories of Aether and Electricity, vol. 2: The Modern Theories, 1900–1926* (London: Thomas Nelson and Sons Ltd., 1953), 39.

146. Hessen, "Ether," 118.

147. Ibid.

148. A. Einstein, "Ether and The Theory of Relativity," in *The Collected Papers of Albert Einstein, vol. 7: The Berlin Years – Writings, 1918–1921*, ed. A. Engel and E. Schucking (Princeton: Princeton University Press, 2002 [1920]), 181.

149. Hessen, "Ether," 118–119.

150. Engels, *Dialectics of Nature*, 523. See also Hessen, "Ether," 118.

151. Hessen, "Ether," 118.

Chapter Four

A Pantheon of Great Ideas

For this final chapter,[1] I will enumerate the ways in which Boris Hessen's *Manuscripts and Documents on the History of Physics* (hereafter, *Manuscripts*) bring our understanding of Hessen's thought full circle, insofar as this text reinforces the overall coherence of his system. A much-fabled work, the composition and dissemination of the *Manuscripts*[2] remain something of a mystery today. Hessen appears to have commenced working on them in the late 1920s and to have completed them sometime between 1934 and 1936. All in all, it consists of Hessen's prefaces and commentaries, as well as primary and secondary sources, that present a dialectical materialist reading of the history of science, with special attention being paid to early modern mechanics. It is believed to have been part of a larger effort to prepare anthologies on the history of science for the Department of the Dialectics of Natural Science at Moscow State University. As far as we know, the original manuscript has been either lost or destroyed, but the layout of the printed version of the work did survive.[3] The text most likely never saw release because of Hessen's fate, though it was slated for publication in 1936 by the [State] United Scientific and Technical Publishing House, which eventually ceased operations in 1938. It is unclear why the layout was only censored and not destroyed, but whatever the case may be, the physical (and the eventual digital) copy of the layout has been preserved and passed down for nearly a century, accumulating careful attention and study along the way, with an official release along with a partial English translation.

In this chapter, I contend that Hessen's *Manuscripts* help us to better understand the overall unity of Hessen's oeuvre, given that they not only show more expansive forms of the arguments that he advanced in the 1931 Newton paper but also echoes themes from throughout the entirety of his oeuvre. This confirms the increasing consensus that while Hessen certainly does emphasize different themes throughout his work, his works seem to be characterized by a remarkable consistency. In the first section, *Sonntagslektüre* or Pantheon of Great Ideas, I lay out Hessen's argument in favor of treating contemporary scientific problems in light of the history of science, whereas the second section, The Metaphysics of Empiricism, provides a reconstruction of Hessen's critique of an empiricist approach to the history and philosophy of science. From there I will move to The Social and Economic Roots of the Scientific Revolution in the third section, in which I will explain Hessen's argument in the *Manuscripts* for the root of classical physics in the socioeconomic development and technological progress engendered by early capitalism, while drawing parallels and divergences with his other works. Finally, the fourth section, A Marxist Classification of the Sciences, summarizes Hessen's Deborinite dialectical materialist approach to the history and philosophy of science as a comprehensive worldview. In support of my claims, I will draw from Hessen's oeuvre, especially the *Manuscripts* and from the 1931 Newton paper, along with selections from the secondary literature.

SONNTAGSLEKTÜRE OR PANTHEON OF GREAT IDEAS

In this first section, I outline the way in which the *Manuscripts* help bring to light one of the central themes of Hessen's entire oeuvre, namely, the importance of the history of science for present-day scientific research. In the Preface to the text, Hessen evokes the following question: Is the history of science a collection of mere artifacts or does it somehow inform the present? To stage the former view, he recounts a private conversation between himself and a German philosopher of science as well as leading figures from the logical empiricist movement, Hans Reichenbach during the Sixth Congress of German Physicists in Königsberg, Germany (today Kaliningrad, Russia). Hessen quotes Reichenbach as having said the following: "[f]or you [Hessen] the history of science is the key to knowledge about its real condition. For us, this is merely *Sonntagslektüre* [entertaining reading on a Sunday]; it cannot provide us with any real knowledge about today's situation in science."[4] In other words, for Reichenbach, while the history of science may provide compelling information, it ultimately bears little to no significance for addressing problems in contemporary scientific research; this was a view which, as Hessen points out, was a widespread sentiment in the early twentieth century and had culminated in neo-positivism.[5] To stage the latter view, he points out that this a-historicism had not always been the prevailing belief, and in fact, as recently as the nineteenth century, scientists had taken for granted the idea that the history of science was indispensable for grappling with the scientific problems of the day:

> [n]early all of the most significant representatives of nineteenth-century classical physics (Ampere, Faraday, Maxwell, von Helmholtz, Boltzmann, Calvin) showed, to some degree, a keen interest in the history of their science and addressed this history not only in particular articles and research, but also wove historical analyses of the issue into their main works.[6]

Hessen laments the former view, not only because he believes that the history of science has intrinsic value but also because it is impossible to understand contemporary science in a truly scientific manner without it. To properly understand the present moment of scientific research in such a way means first and foremost recognizing that it is itself not an endpoint but a phase in an overall process of historical development. As he writes, "[n]o matter how new and unusual the theories of contemporary physics may be, no matter how radically they differ from the outlook of classical physics, the contemporary stage of development in physics is still a historical phase of its overall development."[7]

By recognizing contemporary scientific research in such a fashion, one grasps a great deal more than just moments in the passage of time. Indeed, it allows for the possibility of detecting patterns throughout history, such that the central problematics of science can be more readily distinguished from those more ephemeral. In this way, the history of science is not simply pleasurable reading for one's own enrichment but is a necessary component of understanding any and all scientific problems. He writes, "[i]n acquainting ourselves with the history of physics, we can see that many fundamental issues had been raised earlier and, in some cases, the correct way of resolving these issues had already been outlined. History is, after all, not a 'list of human errors, but a pantheon of great ideas.'"[8]

However, we should note that it was not enough for Hessen to simply reconstruct a chronology of major figures in the history of science. On the contrary, Hessen shunned more conventional narratives that viewed the history of science as a series of contributions by great minds, insofar as he maintains that this too was an insufficiently scientific approach to the essence of the enterprise. From where, then, does the history of science actually emerge? What actually charters the development of the history of science? To address these questions, Hessen proposes his own objection to the empirical method for addressing these very questions.

THE METAPHYSICS OF EMPIRICISM

For this second section, I want to draw the reader's attention to Hessen's critique of an empiricist approach to both the history of science and to contemporary scientific research in the *Manuscripts*. Of course, the term "empiricism" has many different connotations and its meaning has transformed over the course of history, but as defined in the *Dictionary of the History of Science*, it refers to "[a]n ensemble of theories of explanation, definition and justification to the effect that our concepts or knowledge are derived from or to be explicated (or justified) in terms of sense-experience (or introspection)."[9] While Hessen does not explicitly define empiricism himself, based upon his writings, we can safely assume that he understands it along these same lines and that he considers epistemological standpoints (such as Machism, empirio-criticism, empirio-monism, etc.) to belong to the same family of thought. Throughout his corpus, Hessen's critique of empiricism bears many similarities to his critique of mechanical or metaphysical materialism; namely, that on one hand, it played a tremendous role in liberating natural philosophy from scholasticism, but on the other hand, that it introduced pernicious new shortcomings of its own. As with mechanical philosophy, Hessen never rejected empiricism but, rather, objected to treating empiricism as the philosophical foundation of natural science. It marked an important and necessary moment in the history of the development of natural science and, likewise, marks an important and necessary moment in any specific practice of natural scientific research for gathering data and observations. As he writes elsewhere, "[o]ne thing is the 'economical description' as a general, methodological and *gnoseological* principle, but it is quite another thing for description to be a step in the process of physical research."[10] If one universalizes empiricism as the philosophical foundation of natural science, one falls into the trap of either never developing more robust theoretical accounts, or of unforeseen metaphysical speculation in idealistic and spiritualistic forms. Thus, in the *Manuscripts* he claims that "consistent empiricism leads to theology and mysticism."[11]

Hessen presents this argument numerous times throughout his corpus, but many of these are more descriptive than rigorously argued. This is precisely why Hessen's presentation of critiques of empiricism in his *Manuscripts* is so immensely significant. Though Hessen elaborates only a little more of his own critique in this work, he nevertheless presents the sources from which he draws this critique: G. W. F. Hegel's *The Encyclopaedia Logic* (specifically, the chapter titled On Empiricism), Alexander Herzen's *Letters on the Study of Nature* (specifically, Letter 1 – Empiricism and Idealism), and Friedrich Engels's *Dialectics of Nature* (specifically, the chapter titled Natural Science in the Spirit World). I maintain that by reconstructing the arguments provided in these texts, we may develop a better sense of Hessen's critique of empiricism and can see exactly why he saw it was either too restrictive or too speculative for the history of science and for contemporary scientific research. To do this, I will summarize Hegel's, Herzen's, and Engels's accounts, crystallizing their common themes and then reflecting on the extent to which they contribute to our understanding of Hessen's position.

Hegel's *The Encyclopaedia Logic*

In a chapter from *The Encyclopaedia Logic*, titled On Empiricism, Hegel begins by commending empiricism in light of its revolutionary contribution to the history of philosophy, with its challenge to the abstract and rigid categories of scholasticism. Where scholastic philosophy had struggled to grasp concrete particulars, was capable of only grasping the finite through the infinite, and was highly dependent upon authoritative texts, empiricism was directed precisely at grasping concrete particulars by stressing the fundamental nature of experiential knowledge, thereby breaking with the claustrophobia of authority and attending to one's own perspective.[12] The drama of this break was so considerable, however, that in some ways it concealed the degree to which empiricism imported many new abstractions and shortcomings of its own. Hegel primarily stresses two of these shortcomings. First and foremost, empiricism ignores the degree to which experience is always already infused with certain categories of understanding. He writes that "[e]mpiricism falls into error in analyzing objects if it supposes that it leaves them as they are, for, in fact, it transforms what is concrete into something abstract."[13] He elaborates by claiming that

> [t]he fundamental illusion in scientific empiricism is always that it uses the metaphysical categories of matter, force, as well as those of one, many, universality, and the infinite, etc., and it goes on to draw conclusions, guided by categories of this sort, presupposing and applying the forms of syllogising in the process. It does all this without knowing that it thereby itself contains a metaphysics and is engaged in it, and that it is using those categories and their connections in a totally uncritical and unconscious manner.[14]

In other words, experience is not as immediate as it appears; the moment that one begins to analyze an experienced phenomenon, one has to resort to certain categories (i.e., matter, force, unity, multiplicity, etc.) in order to understand it, and these categories are inescapably informed by metaphysical presuppositions. Second, Hegel claims that despite the major success of empiricism in severing the bonds with scholasticism, it nevertheless fails to sufficiently account for human freedom. Empiricism focuses on how experience of the external world can be used as the basis for making truthful claims, but in so doing, it falls back upon a merely descriptive account of the surrounding world. If one has not yet experienced a phenomenon, then it is simply beyond the purview of empirical inquiry. However, human freedom resides precisely at this point, beyond what is immediately given in experienced reality; it charters that which has yet to be done or seen. Hence, Hegel writes that "[e]mpiricism, is cognizant . . . only of what *is*; it does not know that which only *ought* to be, and for that reason is *not there*."[15] He continues by claiming that "[s]ince for Empiricism this sensible domain is and remains something given, this is a doctrine of unfreedom, for freedom consists precisely in my not having any absolute other over against me, but in my being dependent upon a content that is just myself."[16] It is a philosophy that remains forever adhered to the past and to the present, but not one that is in the position to imagine what could be. In this way, it leaves insufficient space for possibility.

Herzen's *Letters on the Study of Nature*

In many ways, Herzen's Letter 1 – Empiricism and Idealism from his *Letters on the Study of Nature* is a continuation of Hegel's both praise and criticism of empiricism. However, he adds an interesting perspective to these same criticisms and introduces several new ones of his own. Like Hegel, Herzen acknowledges the tremendous debt that natural science owes to empiricism, given its challenge to the cold abstractions of scholastic philosophy and reliance on the experience of concrete particulars, as opposed to abstract categories and authoritative texts.[17] Herzen contends, like Hegel, that despite empiricism's claim to remaining confined to experiential knowledge, in practice it does anything but.[18] He goes on to provide reasons for precisely why this takes place.

First and foremost, he argues that empiricism drifts into abstract speculation because it provides an insufficient account of the mind, given that it largely portrays it as a passive receptacle of facts, rather than as an active entity unto itself. He writes, "[t]here is a tendency [in empiricism] to perceive the mind as a passive receptacle, a kind of mirror which would reflect the given without modifying it, i.e., with all its fortuities and without assimilating it—mechanically unreflectingly."[19] The mind is itself part of the world and plays as much of a role in constituting facts as it does in perceiving them. Second, empiricism overlooks the fact that in order to synthesize facts, one must immediately resort to metaphysical assumptions to assemble them into a coherent set of knowledge, given that facts alone amount to little more than a disjointed catalogue of information. Not even the strictest exercise of logical deduction dictates the next step of cognition unquestioningly.[20] If one does this unknowingly, one can mistake one's theoretical speculations for simple descriptions of one's experience. He writes that

> [f]acts are only a collection of uniform material, and not a living growth, however complete the sum of the various parts. Empiricists, realizing this instinctively, turn to abstract reasoning, thinking thus to grasp the whole part by part. That is how they lose the object, existing in reality, and replace it by abstractions, existing only in the mind.[21]

Third, Herzen argues that empiricism provides an inadequate apparatus for grappling with the unknown.[22] The mind, he maintains, abhors a vacuum, meaning that it will always speculate beyond the realm of immediate experience into that which it does not understand. By refusing to provide a theoretical account of the unknown, it inadvertently becomes mired within it. He writes that "[t]he tendency to go beyond the bounds of empiricism is perfectly natural, for one-sidedness is repugnant to the human spirit."[23] Without a proper account of the unknown, the mind embarks on speculation without even realizing it and drifts ever further into abstraction.[24] Fourth, Herzen does not describe this as merely arbitrary striving, however, but instead as a natural tendency of striving towards human freedom.[25] Hence, his argument captures both the tremendous value of empiricism and also its many limitations.

Engels's *Dialectics of Nature*

Engels's most sustained critique of empiricism can be found in his *Dialectics of Nature*, specifically in a chapter titled Natural Science in the Spirit World. Once again, like Hegel and Herzen, Engels explains how in spite of its profound contribution to grounding the natural sciences, empiricism has always been prone towards venturing into idealistic and spiritualistic metaphysical speculation.[26] He mentions that this has been true of empiricism since its very inception, as he explains how despite Sir Francis Bacon's status as the father of empiricism, Bacon also sought out more esoteric pursuits such as alchemy, performing miracles, creating new species, mastering the weather, and so on.[27] This is no less true of Newton, who despite being the epitomizing figure of empiricism in classical physics became obsessed with the Book of Revelation in his later years, for instance.[28] Engels then proceeds to discuss how this legacy had only continued into the nineteenth century, as he cites numerous contemporary examples of the marriage of empiricism with more esoteric practices such as hypnotism, mediums, photographing spirits, and so on.[29] For Engels, one of the central failures of empiricism is its lack of a sufficient account of speculative theory, he specifies, and of dialectics in particular. One of the central tenets of dialectical philosophy, Engels argues, is that any phenomenon always not only entails its opposite but inevitably becomes its opposite. Hence, it should come as no surprise that a philosophy that insists on adhering so closely to experience, and which denies the validity of dialectical play, eventually unconsciously wanders into the most abstruse types of metaphysical speculation. He writes, "according to an old and well-known dialectical law, incorrect thinking, carried to its logical conclusion, inevitably arrives at the opposite of its point of departure. Hence, the empirical contempt for dialectics is punished by some of the most sober empiricists being led into the most barren of all superstitions, into modern spiritualism."[30] In other words, even as empiricism tries to stand as a bulwark against theoretical speculation, it also embraces them all the more closely.

Analysis: The Metaphysics of Empiricism

Based on Hessen's sources here, then, we can extrapolate that his critique of empiricism can roughly be summarized as follows: The road that empiricism takes to escape philosophical speculation effectively leads it right back to square one in a number of ways. First and foremost, empiricism treats the subject as the passive observer of the object, thereby assuming that the subject grasps the object of perception in its immediacy. Of course, however, this overlooks the fact that the subject is itself a moment of the world. In this way, it fails to recognize what the subject introduces to perception. Secondly, empiricism always already contains a metaphysics from the moment one begins to analyze one's own perceptions. Objects of experience do not immediately explain themselves; this invariably involves speculation into the unknown. Without a sound method for speculating and inferring about the unknown, this becomes an abstraction that merely reify present. Finally, empiricism can only ever remain tethered to the so-called "facts;" that is, it can only provide a summary observation of what has already happened. In this respect, it is unable to provide an account of that which is possible. It overlooks the central role played by horizons of possibility via freedom in the structuring of human knowledge, precisely because the realization of the possible is an essential moment of human inquiry.

THE SOCIAL AND ECONOMIC ROOTS OF THE SCIENTIFIC REVOLUTION

I will now proceed by analyzing Hessen's approach in the *Manuscripts*, showing both its similarities to and notable differences from the 1931 Newton paper. Exegetically speaking, I contend that the *Manuscripts* are an elaboration upon the first two subtheses of the 1931 Newton paper, though with some notable exceptions. Generally speaking, the 1931 Newton paper was designed to illustrate how the socioeconomic development and technological progress of early capitalism cultivated a milieu that was a necessary condition for the composition of a work like Sir Isaac Newton's magnum opus, *Mathematical Principles of Natural Philosophy*. The *Manuscripts*, though affording no less significance to Newton, effectively makes this same argument, but with respect to the entirety of the Scientific Revolution. In this section, I will outline Hessen's main arguments in the *Manuscripts*, noting major similarities with the 1931 Newton paper along the way, along with some of the noteworthy differences that help to bridge the apparent gaps in his thought.

The *Manuscripts* elaborate upon the 1931 Newton paper to the extent that the central thesis of the two works is the same. In fact, they are basically identical, as Hessen argues that "[t]he remarkable flourishing of the natural sciences in the sixteenth and seventeenth centuries is due to the break-up of feudal ownership, and the development of merchant capital, international maritime transport and heavy industry (mining and metallurgy)."[31] In other words, from the breakdown of the feudal economy and the ascent of capitalism as the prevailing mode of production, the transition from a consumption-based economy to an exchange-based economy made for a different *modus operandi* of production and in which machinery began to take on a central function within economic production. These major shifts resulted in the need to theorize about the nature of machinery from a mathematical perspective, thereby laying the groundwork for classical mechanics.

To properly establish his primary thesis, Hessen contends that one cannot understand this process solely chronologically but must, rather, see it from both diachronic and synchronic perspectives. Hessen chooses to take a thematic approach; the three themes of the *Manuscripts* effectively being a reformulation of subtheses 1 and 2 of the 1931 Newton paper. The first thesis of the 1931 Newton paper consists of two moments, "The Economics, Technology and Physics of Newton's Era/The Physical Themes of the Era and the Contents of the Principia," in which the former part shows the ways in which the breakdown of the feudal economy and the dawn of the capitalist economy established new technical demands, while the latter shows how these technical demands constituted the practical basis of what would become the theoretical problems of modern physics, thereby engendering the mechanistic conception of material causality.[32] These two components of the first thesis are transformed into two independent themes in the *Manuscripts*, such that "The Economics, Technology and Physics of Newton's Era" becomes "The Socio-Economic Prerequisites for the Emergence of Classical Physics" and where "The Physical Themes of the Era and the Contents of the *Principia*" becomes "The Emergence and Development of the Main Principles of Classical Mechanics and the Arguments surrounding Them in the 17th Century." The second thesis of the 1931 Newton paper, "The Class Struggle during the English Revolution and Newton's Worldview," is designed to show how the mechanistic conception of materialist causality provided an inadequate conception of the physical world and how natural philosophers and scientists invariably debated about the nature of divine or spiritual causes; these divine or spiritual causes were frequently rationalizations of religious beliefs among classes of thinkers.[33] This thesis effectively becomes "The Struggle of Materialism and Idealism concerning This Problem in the 17th and 18th Centuries." We should note that there are also numerous places in which Hessen extracts lines essentially verbatim from the 1931 text. To do so, he presents primary sources in the history of physics, primary sources in the socioeconomic history, secondary sources in the history of physics and socioeconomic history, all of which he narrates with his own prefaces and commentaries.

In Theme 1, "Socio-Economic Prerequisites for the Emergence of Classical Physics," Hessen explains how the breakdown of the feudal economy and the emergence of capitalism were the necessary condition for the rise of modern physics. He states, "the first theme [is] aimed at showing how the break-up of feudalism and the development of a new mode of production brought a number of new technical tasks into existence and presented physics with a series of issues predominantly of a dynamic character."[34] This theme is broken up into four subthemes that can roughly be grouped into four categories: trade and transport, warfare, mining and metallurgy, and engineering. Across these different subthemes, Hessen shows how the dawn of the capitalist economy radically reshaped both the nature of the physical world as well as the ways in which we perceive it. To the extent that economic production was increasingly being governed by the merchant class and production was being organized for the sake of exchange, this created an increasing demand for improving trade routes by land and by sea. Where feudal lords had an incentive to retain shabby roads, given the *Grundruhrecht*, according to which lords had the right to any goods that fell on their land, it became necessary to improve and construct new roads with the dawn of capitalism. Likewise, this created a demand to improve river and sea transport. In addition to improving shipbuilding to complete such journeys, this also led to the development of canal and lock construction along with the innovation of the pendulum clock to keep time while traveling at sea. Next, the development of capitalism coincided with the growth of powerful nation states and competition for foreign markets, thereby accelerating the production of arms/artillery along with fortifications. Furthermore, the increased circulation of money required the greater extraction of copper, gold, iron and silver, thereby leading to improvements in mining. Finally, the new world established the conditions for the emergence and predominance of engineering as a profession. In his summaries of each of these subthemes, Hessen consolidates the practical challenges and the theoretical questions that emerged therefrom.[35]

Theme 2, "The Emergence and Development of the Main Principles of Classical Mechanics and the Arguments Surrounding Them in the 17th Century," explains how these various practical and theoretical tasks were consolidated into general theories of matter and motion. As Hessen points out, "[t]his section provides a survey of the emergence and development of the main principles of dynamics."[36] Though the shortest of the three themes, Theme 2 primarily focuses on the way in which physicists' immersion in the practical tasks mentioned in Theme 1 provided the basis of their theoretical speculation. Likewise, Hessen points out that the domain of modern theoretical physics is not a static terrain but is, rather, one infused with central tensions between major figures such as, for instance, Descartes, Huygens, Leibniz, and Newton, as well as between major concepts in physics regarding the conception and measurement of both matter and motion (e.g., is matter active or passive, and is motion absolute or relative?). Thus, in addition to the socioeconomic preconditions, Hessen shows the central tensions between different concepts animated the development of physics as well.[37]

Theme 3, "The Struggle of Materialism and Idealism concerning This Problem in the 17th and 18th Centuries," shows the different topics that fall under, what could be called in Marxist parlance, "ideological struggle." Here, Hessen addresses effectively two dimensions of this ideological struggle: first and foremost, between feudal universities and scientific societies and second, the ways in which the limitations of the mechanistic conception of material causation left the door open to incorporating immaterial causes and, therefore, provided ways of defending against atheism and materialism. To elaborate upon the first point, Hessen explains how the universities were effectively the bastion of feudal society, to the extent that they produced the elite of the day and, through their studies, inculcated all scientific study with the theological principles and moral values of the Church. In this respect, they resisted the efforts of the emerging bourgeois class to promote their scientific studies, meaning that rather than becoming a part of the university system, they had to form their own societies. The most famous of these scientific societies were the Florentine Academy del Cimento, the London Royal Society, and the Parisian Academy of Sciences. Such societies provided the ascendant merchant class with a means for institutionalizing their interests; namely, to gather in order to share knowledge on economic and technical matters and to disseminate such knowledge through their own publications. Hessen also notes, however, that despite the fact that these societies remained steadfast in their commitment to the new scientific worldview, debates arose about the relationship between material and spiritual or divine causation because of both the limitations inherent to empiricism and to the mechanistic conception of material causation and the resistance to fully fledged atheism.[38]

I maintain that the *Manuscripts* confirm an argument that many Hessen scholars have long been making; namely, that the 1931 Newton paper was never nearly as simplistic of a work as it was often made out to be.[39] While many have already detected this in the 1931 text itself or through the increased study of Hessen's other works, the *Manuscripts* only further confirms this in a number of different ways. First and foremost, the *Manuscripts* confound the notion that the 1931 Newton paper was some kind of rushed exception to Hessen's overall project. By reconstructing the chronology of Hessen's composition of the *Manuscripts*, we can see that it appears to have been a project on which Hessen had been working since the late 1920s, at a time during which he was composing those works which are often considered more internalist in nature. That Hessen was composing these works side by side substantially mitigates the notion that he underwent some distinct break in his thinking. Second, the actual contents of the *Manuscripts* help to bridge the gap between the 1931 Newton paper and the rest of Hessen's oeuvre. In the *Manuscripts*, however, we see Hessen placing both positions side by side. While subtheses 1 and 2 roughly correlate to themes 1 and 2, in theme 2, Hessen, in mentioning a wider swath of thinkers, not only presents excerpts from classical works in modern physics but also isolates the major dialectical tensions between the systems of different physicists that propelled the theoretical debate forward. Third, while the *Manuscripts* certainly confirm that Hessen saw socioeconomic factors as foundational to the birth of modern physics, the sheer length of the work allows Hessen's vision greater room to breathe. In this expanded form, we see Hessen presenting a picture of scientific work that is far less schematic and clean. This sense is perhaps furthered by the fact that the style of presentation is so different between the 1931 text and the *Manuscripts*. Even though Hessen narrates his standpoint in prefaces and narration, by including other authorial voices via lengthy quotations of primary and secondary sources in physics, sources in history, and so on, most of which do not align with Hessen's ideological commitments, the reader both gets Hessen's interpretation of how socioeconomic forces as well as other authors' conceptions of what animates these transformations. This lends itself to a greater sense of the complexity and, to some degree, the contingency of how certain practical and theoretical developments ensued.

There are, likewise, several noteworthy differences between the two texts that invite further analysis. First and foremost, as alluded to previously, the

Manuscripts do not incorporate any discussion of subtheses 3 and 4 from the 1931 Newton paper. Subthesis 3, "Engels' Conception of Energy and the Lack of the Law of Conservation of Energy in Newton," refers to the upper limit of progress of early modern physics; that is, despite the genius of figures like Newton, they did not notice certain matters that would eventually become obvious in the discussion of thermodynamics. Hessen contends that this was not due to the fact that they were unintelligent but was, instead, due to the fact that the technological basis had yet to be furnished. Thermodynamics was not possible until the steam engine began to play a central role in economic production during the second Industrial Revolution.[40] Subthesis 4, "The Machine-Wreckers in Newton's Age and the Present-Day Wreckers of the Productive Forces," refers to the way in which machine-wrecking was a manifestation of workers enacting protest, while still imbued with bourgeois ideology which fixated on technology, rather than directing their attention to challenging class exploitation. The point here is to discuss how a transformation of socioeconomic relations would usher in a new perspective that would invariably construct the basis for a new type of science.[41] Once again, Hessen makes no mention of these topics in the *Manuscripts*. Another of the fascinating differences between the *Manuscripts* and the 1931 Newton paper is that Hessen's discussion of the struggle between the old university system and scientific societies is that the former treats it as part of the ideological struggle, while the latter treats it as part of the socioeconomic roots. Next, by virtue of the length, there are numerous historical details in the *Manuscripts* that are simply not mentioned in the 1931 piece (i.e., the deliberate sabotaging of roads, the design of fifth wheels on carriages, clocks, scientific societies beyond the London Royal Society, etc.). Perhaps one of the more compelling is Hessen's inclusion of a reference to Johannes Mathesius's *Sarepta*. This work incorporates a fascinating detail not mentioned at all in the 1931 Newton paper, namely, the way in which Protestantism was not simply a way of "eternalizing" the capitalist order but, more concretely, of furnishing the religious sentiment that supplanted the ideological foundation for labor under capitalism. Take, for instance, Mathesius's line in which he states that "God's mercy and gift is that, with the help of useful contraptions and tools, heavy labor carried out by the sweat of one's brow, imposed upon humankind for its sins, is eased."[42] This supplants my prior claim about Hessen seeing the importance of ideology within the domain of practice, along with recognizing the importance of ideological struggle in the realm of theory. Surely, further study of the 1931 paper and the *Manuscripts* side by side will reveal still other subtle and noteworthy differences. As it stands, however, it is a key document for showing the more expansive thought process behind the 1931 Newton paper but which also proves to be entirely in sync with his other work.

A MARXIST CLASSIFICATION OF THE SCIENCES

In this final section, I would like to provide a synthesis of the major points of Hessen's Deborinite dialectical materialist philosophy. As indicated, it would be folly to assume that dialectical materialism was only ever one thing. Debates over what dialectical materialism actually was and how it should be applied were widespread throughout much of Hessen's lifetime. One could not go so far as to call dialectical materialism an empty signifier that meant whatever to whomever, but there were undoubtedly distinct camps of dialectical materialist philosophers, along with distinct individual flourishes to this way of thinking, by which one could hardly call them the same thing. Based upon these premises, Hessen's articulation of dialectical materialism can be best encapsulated under the following parameters: First, dialectical materialism is a fundamental ontology, and second, it is an epistemological standpoint, all of which are designed to establish the possibility of ushering in a new society. For this concluding section, I will draw from works written throughout the entirety of Hessen's oeuvre.

First and foremost, Hessen saw dialectical materialism as a fundamental ontology (i.e., a philosophy concerning the essence of reality itself). As he writes, "[s]tudying the real processes of the motion of matter and human knowledge leads us to believe that dialectical contradiction is the objective contradiction of all motion, and that 'dialectics reduced itself to the science of the general laws of motion, both of the external world and of human society [*sic*]' (Marx)."[43] By materialist, Hessen means that reality is a mind-independent, extended substratum and by dialectical, he means that reality is defined by self-motion that is animated through the unity of opposites. In support of the former, he writes that reality is an "(extended) substance, objectively existing regardless of us,"[44] while in support of the latter, he writes that "dialectics teaches us to consider each object in its development, its self-movement; to study the immanent laws of the development of the object. . . . The unity of opposites, the inconsistency of all the connections and relations of the object is the basis of its self-movement."[45] While the synthesis of the terms may appear somewhat strange, we can perhaps better understand it by framing it as an "anti-Platonic" position. Recall that, according to a rather elementary reading of Plato, physical reality is defined by contradictions, and it is because of these contradictions that the truth must lie in a transcendent realm of self-identical forms.[46] For a dialectical materialist, however, precisely because there is no transcendent realm, the contradictory nature of the physical world does not convey its deficiency as reality but instead as reality itself. One of the outcomes of this position is the argument that human nature must be understood from the perspective not of ideas but of material praxis (i.e., the social relations and technological forms from which human beings sustain themselves and shape their reality across history).[47] As mentioned previously, he saw it as an antireductionist natural philosophy; Hessen's own antireductionism being exemplified in the following quote, in which he writes that "[w]e cannot reduce or dissolve a higher form of motion into the sum of the lower forms of motion, since the specificity of the higher form of motion lies precisely in the fact that it represents not the sum of the lower forms, but their synthesis."[48]

Second, Hessen treats dialectical materialism not only as a fundamental ontology but also as an epistemological standpoint. This epistemological standpoint entails not only explaining reality in dialectical materialist terms but also of reflecting upon the condition of the possibility of explaining reality in such terms. Hessen identifies this as a two-fold program that he calls a "Marxist classification of the sciences." This program consists on the one hand of a "study of the historical sequence of the forms of motion as they appear in the development of the science of physics in human society," and on the other, as a "study [of] the 'natural science of the development of matter."[49] In other words, dialectical materialism explains how the perception of reality is rooted in socioeconomic development and technological progress, while also understanding that scientific theories unfold according to competing understandings of matter and motion.

In sum, then, the sufficient condition for the emergence of natural scientific theories depends not only upon a historical standpoint but also on a standpoint that is grounded in dialectical materialist philosophy. Scientific theories independent of history can neither scientifically explain their own emergence, nor can they provide a conventional history of science. On the contrary, as Hessen argues, a scientific approach to scientific theories entails the assumption of an immanent, self-moving material reality and, as such, of human nature being fundamentally animated by socioeconomic development and technological progress, which is inherently historical. The emergence of scientific theories, then, requires casting them in the broader milieu of socioeconomic development and technological progress, along with seeing the historical development of ideas in natural philosophical and scientific theory that must be grasped as competing theories in the development of the understanding of both matter and motion.

For Hessen, dialectical materialist philosophy as a fundamental ontology and epistemology was the key to navigating the long road to constructing a classless society. He saw in the perceived breakdown of capitalism and the perceived dawn of socialism the basis of a new science just as the breakdown of feudalism and the dawn of early capitalism had furnished the basis of classical mechanics via socioeconomic development and technological progress. Likewise, just as socioeconomic development and technological progress alone left certain physical questions unanswered, questions which were dealt with via ideological struggle, the same would be the case for both quantum mechanics and relativity theory. Just as the merchant class had formed scientific societies in the ideological struggle over the new science during the rise of capitalism, so too the working class would have to form their own institutions in the ideological struggle over a new science during the expected rise of socialism. This involved not completely objecting to the study of quantum mechanics and relativity theory but, rather, cultivating their study in the context of working-class scientific institutions facilitated by cooperation with experts in their respective scientific fields.[50] This, he maintained, would be the path to being and knowing at the gates to the new world.

NOTES

1. This chapter is a revised version of my “A Pantheon of Great Ideas: Boris Hessen and the History and Philosophy of Science,” in *Manuscripts and Documents on the History of Physics: A Historical Materialist Textbook*, ed. Pietro Omodeo and Sean Winkler (Venice: Verum Factum, 2022), 35–73.

2. Even though the title page and the original table of contents of the work are missing (presumably censored), we do know that Hessen meant for it to be titled *Manuscripts and Documents on the History of Physics*, based on the footer of p. 385. В. С. Кирсанов, *Избранные труды. Воспоманния коллег и друзей. Стики. Ризунски*, ed. Д.А. Баюк (Москва: Издательство им. Сабашниковых, 2010), 98–99.

3. Ibid., 97.

4. Hessen, *Manuscripts and Documents on the History of Physics*, 187.

5. Ibid.

6. Ibid.

7. Ibid.

8. Ibid.

9. R. Bhaskar, “Empiricism,” in *Dictionary of the History of Science*, ed. W. F. Bynum, E. J. Browne and R. Porter (London: The Macmillan Press Ltd., 1981), 121.

10. Hessen and Egorshin, “On Cde. Timiryazev’s Attitude towards Contemporary Science,” 146.

11. Hessen, *Manuscripts and Documents on the History of Physics*, 231.

12. G. W. F. Hegel, *The Encyclopaedia Logic (with the Zusätze), Pt.1*, trans. T. F. Geraets, W. A. Suchting and H. S. Harris (Indianapolis: Hackett Publishing Company, Inc., 1991), 76, 77.
13. Ibid., 78.
14. Ibid., 77–78.
15. Ibid., 77.
16. Ibid., 79.
17. A. Herzen, "Letter 1: Empiricism and Idealism," in *Selected Philosophical Works*, trans. L. Navrozov (Moscow: Foreign Languages Publishing House, 1946), 104, 107, 117–18.
18. Ibid., 107.
19. Ibid., 111.
20. Ibid., 107.
21. Ibid., 109.
22. Ibid., 101.
23. Ibid., 108.
24. Ibid.
25. Ibid., 108–9.
26. Engels, *Dialectics of Nature*, 345.
27. Ibid.
28. Ibid.
29. Ibid., 345–355.
30. Ibid., 354.
31. Hessen, *Manuscripts and Documents on the History of Physics*, 190. See also Hessen, "The Social and Economic Roots of Newton's *Principia*," 44.
32. Hessen, "The Social and Economic Roots of Newton's *Principia*," 43–52, 52–61.
33. Ibid., 61–73.
34. Hessen, *Manuscripts and Documents on the History of Physics*, 190.
35. Ibid., 190–223.
36. Ibid., 224.
37. Ibid., 224–229.
38. Ibid., 230–266.
39. Cohen, *The Scientific Revolution*, 332. See also Freudenthal, "The Hessen-Grossmann Thesis," 167; Freudenthal and McLaughlin, "Classical Marxist Historiography of Science," 32.
40. Hessen, "The Social and Economic Roots of Newton's *Principia*," 73–82.
41. Ibid., 82–89.
42. Hessen, *Manuscripts and Documents on the History of Physics*, 216. See also J. Mathesisus, *Sarepta oder Bergpostill; sampt der Jochimßthalischen kurtzen Chroniken* (Nürnberg, 1562).
43. Hessen, "Materialist Dialectics and Modern Physics," 240–41.
44. Гессен, «Предисловие к статья Г. Мие «Проблема материи»», 119.

45. Гессен and Подволоцкий, «Философские корни правого оппортунизма», 3. See also Hessen, "On the Question of the Causality Problem in Quantum Mechanics," 142.

46. Plato, *Republic*, in *Plato: Complete Works*, ed. J. M. Cooper and trans. G. M. A. Grube and C. D. C. Reeve (Indianapolis: Hackett Publishing Company, 1997), 1132–55.

47. Hessen, "Materialist Dialectics and Modern Physics," 240–41.

48. Hessen, "Mechanical Materialism and Modern Physics," 173.

49. Hessen, "The Social and Economic Roots of Newton's *Principia*," 80.

50. See Hessen and Egorshin, "On Cde. Timiryazev's Attitude towards Contemporary Science," 152; Б. М. Гессен and С. И. Луппол, «О кружках по изучению диалектического материализма среди молодых научных работников», *Коммунистическая революция* 14 (1928): 77–83; Б. М. Гессен, «К вопросу о подготовке научной смены в области теоретического естествознания», Коммунистическая революция 5 (1929): 62–67.

Conclusion

My aim in this book was to provide the reader with an introduction to Soviet philosopher, Boris Hessen's approach to the history and philosophy of science. Because the study of Hessen's thought has until recently remained largely restricted to his "The Social and Economic Roots of Newton's *Principia*" (the 1931 Newton paper), I hoped to more extensively engage with his oeuvre and provide a better sense of the philosophical underpinnings of his work. Through this study, I developed the thesis that Hessen's thought is marked by a consistent defense of a Deborinite dialectical materialist approach to the history and philosophy of science. I began by presenting Hessen's implicit understanding of practice and ideology; two of the central concepts of his system. This was followed by an analysis of his understanding of how ideological conflict, particularly between idealistic and materialistic positions within philosophy and the sciences, foster the ongoing development of scientific theories. I then presented a reading of Hessen's effort to apply dialectical materialism to modern physics, via quantum mechanics and relativity theory, and then concluded with a study of the cohesiveness of his system as a whole. I constructed this position through careful study of Hessen's major works, including his 1931 Newton paper, contributions to *Under the Banner of Marxism*, Preface to the Russian translation of Arthur Haas's *Wave Mechanics and the New Quantum Theory*, *The Main Ideas of Relativity Theory*, contributions to the *Soviet Encyclopedia* and *Manuscripts and Documents on the History of Physics*. Though Hessen's oeuvre stands as one of the most complex and nuanced efforts to apply a dialectical materialist approach to the history and philosophy of science, the conclusion of this work nevertheless brings us to the following question: What value is there in entertaining this philosophy and why should it not simply be cast into the proverbial dustbin of history? I maintain that we can appreciate Hessen's thought from both a historical and a contemporary perspective.

From a historical standpoint, we can appreciate Hessen by virtue of the context in which he wrote. While it may be difficult to appreciate now, the October Revolution of 1917 did mark a singular event in the history of the world which initially marked the prospect of a genuinely transformative social order. Prior to what it became, the revolution was a time that saw the working women and men of Russia rise up in the "ten days that shook the world,"[1] and more than realizing the aims of better labor conditions, hours and wages, thousands hung their hopes on this first effort that the world had ever seen to overthrow the decrepit institutions of the old world and to condemn exploitation, poverty, and servitude as things of the past. While it did not take long for this dream to become perverted, in the late 1920s, some believed that the fate of the nascent Soviet state was still undecided. Recovering from the ravages of civil war, enduring its isolated status on the international stage, and already enacting severe mechanisms of repression, hope may have appeared dim but, arguably, not yet entirely lost for an emancipatory future. This placed the nation at a crossroads of a confounding number of practical and theoretical questions. Hessen was undoubtedly one of the most compelling theoreticians of this immediate postrevolutionary era and a fascinating example of someone trying to remain faithful to a political project predicated on universal emancipation, while recognizing the ways in which the parameters of such a project would threaten to crush it from within if it did not retain an openness of thought. As the official state ideology of the Soviet Union, of course, dialectical materialism would go on to become an instrument of ideological indoctrination and at times, a pretext for exercising state terror. At the level of research, not only was this ideology an impediment to scientific progress, but also in some cases, it was even a pretext to catastrophe, as in the infamous "Lysenko affair."[2] And while the story of the Soviet experiment is undoubtedly a complex one marked by some genuine achievements, none of which can be adequately addressed in a single volume, let alone a single conclusion, what achievements it did make were often at a tremendous cost of human and animal life, suffering, civil liberties, and to the environment. In reading Hessen's approach, one can only wonder whether his thought would have marked a more prosperous path not taken or not.

From a contemporary standpoint, one can see both the blind spots and also the enduring insights of Hessen's work. Despite the complexities of his thought, one can still find in it the substantial limitations of dialectical materialism. While entertaining a greater openness to non-Marxist philosophies, Hessen still insists on this idea that philosophy is always already a form of partisanship. Even though there is without question a profound need to scrutinize philosophies of the past along the lines of not only class but also of gendered and racial bias, dialectical materialism even in its most nuanced forms perpetually seems to treat other philosophies that do not share its own version of emancipation as enemies; Hessen's work is not entirely immune from this reflex. Moreover, one cannot help but get the sense from reading Hessen's work that the effort to entertain complexity of thought perpetually threaten to burst the limits of his dialectical materialist framework. Time and again, one sees the only meaningful difference within dialectical materialism as being that of opposition. So, even when appreciating the nuances of Newtonian versus Cartesian mechanics, or quantum mechanics versus relativity theory, there is a degree to which these theories always must be simplified to fit a particular schema, leaving all too little room for ambiguity, error, and so on. But, it would be all too easy to cast aside Hessen's work as though it were simply antiquated through and through. In its reconstruction of the complex socioeconomic conditions and forms of technological progress that ushered in classical and modern physics, he offers a way of understanding how social conditions play a massively important role in the construction of scientific objectivity. One need not entertain Hessen's politics to recognize the ongoing relevance of this approach to understanding such issues as climate change, the COVID-19 pandemic, and so on. It can also offer a profound counterpoint to the increasingly frequent marshaling of empiricism by reactionary politics, which so often argues that because inequalities and injustices exists that they are written in stone. While this would certainly not address every contributing factor to the challenges we face, and would not suffice without pursuing other alternatives, using an argument akin to Hessen's, it would mean that different potencies of scientific development would be pursued.

Perhaps, there is one final lesson to take away from this study of Hessen's thought. In this time of seeking new alternatives, I would suggest that it is the idea that philosophical questions can be answered. One could very well argue that philosophical questions precisely cannot be answered at all and that their answers are merely part of the endless conversation of humanity. To treat a philosophical question as answered means to deny something that belongs to humanity to forever ponder. And while the crises of the day undoubtedly call for profound new alternatives, our ability to enact them depends on finding some deeper premise on which we can cooperate. Perhaps, whatever that saving grace might be, it will not come in the form of definitive answers but of our shared efforts to wrestle with the most undying perplexities of the world.

NOTES

1. See J. Reed, *Ten Days that Shook the World* (New York: Penguin Classics, 2007 [1920]).

2. See D. Lecourt, *Proletarian Science?: The Case of Lysenko*, trans. B. Brewster (Atlantic Highlands: Humanities Press, 2003 [1977]).

Postscript

Boris Hessen's Entries to the Soviet Encyclopedia

This Postscript is dedicated to illuminating the background of Boris Hessen's contributions to the *Soviet Encyclopedia*. From 1930 to 1933, Hessen contributed six entries to the *Soviet Encyclopedias*, *Great* and *Small*, in which he elaborated upon a series of topics from contemporary physics: "Space,"[1] "Ether,"[2] "Energy,"[3] "Entropy,"[4] "The Ergodic Hypothesis,"[5] and "Einstein"[6] (the last of which was coauthored). While the texts are stand-alone introductions to various scientific concepts and figures, they provide one of the most compelling portraits of Hessen's application of Deborinite dialectical materialist philosophy to contemporary scientific questions when examined alongside one another. Even though they were part of the closely monitored, state-sanctioned endeavor of the *Soviet Encyclopedia*, a better understanding of the circumstances of their production and Hessen's role therein can help us to better see how Hessen's philosophy developed. They also provide an opportunity to see what his philosophical positions actually were. While other scholars have certainly mentioned these entries and have shed valuable light on their impact on Hessen's career trajectory,[7] none have explicitly dealt with the content of these works.

Hessen wrote all six of his contributions to the *Soviet Encyclopedia* for the first edition of the project; five to the *Great*—volumes 65 (1931), 64 (1933), and 63 (1933)—and one to the *Small*—volume 6 (1931). The unusual publication schedule of the project (detailed herein) meant that volume 65 of the *Great Soviet Encyclopedia*, which contains Hessen's entry on "Ether," was released on September 1, 1931;[8] volume 64, containing his entries on "Energy," "Entropy," and "The Ergodic Hypothesis" were completed on October 20, 1932, but were not published until September 18, 1933;[9] and volume 63, which contains his coauthored piece on "Einstein," was completed on February 1, 1933 but was not published until October 1, 1933.[10] Lastly, volume 6 of the *Small Soviet Encyclopedia*, containing Hessen's entry on "Space," was completed on April 29, 1930, and published without changes in January 1931.[11] From their titles, these works would appear to be little more than conventional entries, such as might be found in any scientific encyclopedia. In the context of the *Soviet Encyclopedia* project, however, their story is considerably more complex; each piece bears the imprint of two basic aims that were sought out by Soviet leadership: first, of constructing a socialist society and second, of being at the forefront of the scientific and technological achievements of the day.[12] At times, these aims were complementary and at others, they were wholly incompatible, but in either case, they were the ever-present forces that imbued the individual contributions with their own unique character. In this Postscript, I will attempt to introduce the translations of Hessen's entries by situating them within the larger context of the production of the first edition of the *Soviet Encyclopedia*, showing that they cannot be understood independently of this central tension that surrounds the aims, production, editing, and dissemination of this grandiose project.

The *Soviet Encyclopedia* was the first endeavor to construct a comprehensive knowledge based on the foundation of the dialectical materialist worldview. To this day, it remains not only the grandest and most ambitious endeavor of its kind but also one of the boldest encyclopedic undertakings the world has ever seen. The scholar Paul Josephson aptly refers to it as the "Soviet Britannica."[13] The project was first envisioned by Soviet scientist and explorer, Otto Schmidt, and officially began when the Central Committee of the Communist Party of the Soviet Union (CPSU) approved his formal petition on April 17, 1924, and named him its chief editor on January 15, 1925.[14] Schmidt disputed the CPSU's aim of treating the encyclopedia as a propaganda tool domestically and internationally, but they nevertheless agreed that the Tsarist era *Brockhaus and Efron Encyclopedic Dictionary* (1890–1904) and *Granat Encyclopedic Dictionary* (1910–1917) were outmoded artifacts of the old "capitalist" worldview that did not measure up to the ambitions of the "socialist" endeavor.[15] To accomplish this aim, the encyclopedia aimed to provide a comprehensive account of the world as seen through the lens of dialectical materialism, showing both the promise for such a project to bring about a new world and for it to extend new horizons of human understanding. This would mean espousing a clear political program and cutting-edge knowledge, all in an easily affordable form, in keeping with the values of their politics.[16] As we will see, the interplay of political and intellectual ambitions for the encyclopedia was both the source of its promise as much as it was of its limitations. It would eventually see release in three editions over the near century's long existence of the Soviet Union: first 1926–1947 (sixty-five volumes); second, 1950–1958 (fifty volumes); and third, 1969–1978 (thirty volumes).[17] Upon the dissolution of the USSR in 1991, the project was eventually revived and considerably revised to become the *Great Russian Encyclopedia*, the production of which continues to this day.[18]

The realization of the aims of the project played out in considerably different ways, depending on what the conditions of Soviet life were at any given moment; these conditions were particularly volatile in the early decades of the union. The *Soviet Encyclopedia* originated in the unique time of the NEP (1921–1928), which has been mentioned previously. A much-needed relief from the dire years of War Communism (1918–1921), the NEP introduced a mixed economy in which the state retained ownership over matters such as banking, heavy industry, and trade but otherwise permitted a more tempered approach to private industry. This was a time marked by at least some relative liberalization in the academic sphere, especially for engineers and scientists, who were allowed to pursue research more freely and on their own terms. Likewise, private publishing houses could reopen, providing a much-needed outlet for non-Marxist authors, whose works often went unpublished by the State Publishing House or in journals like *Under the Banner of Marxism*. This liberalization proved to be beneficial not only to non-Marxist theoreticians but also to Marxists because it fostered fruitful intellectual dialogue in many cases. The *Soviet Encyclopedia* of these early years included pieces authored by theoreticians from across the political spectrum in part because of these aforementioned dialogues.[19] As such, it stands out as the most compelling and promising version of the project overall, given that it opened compelling, new questions in many theoretical endeavors. As scholar Loren Graham argues, "[the first] edition of the major Soviet reference work will always be regarded as the classic one of the revolutionary period, retaining interest even yet as a novel effort to present an alternative form of knowledge on all cultural and intellectual issues. The volumes that were published before 1930 are the most interesting."[20]

This would all drastically change in 1929, after Joseph Stalin declared the end of the NEP to launch the Great Break, also previously mentioned.[21] Exemplified in his article, "A Year of Great Change,"[22] the Great Break marked an effort to wholly eliminate private ownership, to accelerate heavy industry and to force the collectivization of agriculture, at often untold human and environmental costs. The CPSU embarked on an effort to "proletarianize" all academic spheres; that is, to suppress any research that was not expressly directed toward the construction of a socialist society. This brought an abrupt shift to the previous years' intellectual debates, as Stalin deemed this theoretical openness to come at the detriment of concrete practice.[23] As a product of the NEP, the *Soviet Encyclopedia* immediately fell into the crosshairs, particularly the Deborinite branch of the dialectical materialists, who were often contributors and editors and who had embraced the more nuanced approach to non-Marxist viewpoints of the NEP years.[24] In December 1930, Stalin even personally accused the editorial board of the encyclopedia of being too relaxed with so-called "deviationists" and referred to the theoretical vision of the project as "'at the very least a muddle."[25] Later that month, the CPSU cell of the Institute of Red Professors alleged that the Deborinites were even smuggling anti-Marxist rhetoric into the volumes[26] and, in July 1931, placed production of the encyclopedia under the supervision of the head of Kultprop.[27] That year saw a massive purge of the editorial board, including Deborin himself,[28] such that within a matter of months all of the encyclopedia's editors and 80 percent of the subsection editors were replaced by party members.[29] Graham writes that "the later ones fell under the Stalinist controls that spread throughout Soviet life, extinguishing the spirit of innovation and experimentation that had been present in the Soviet cultural and intellectual worlds in the twenties."[30]

The foundational aims of the *Soviet Encyclopedia*, combined with the drastic circumstantial shifts mentioned, led to some rather paradoxical outcomes. In theory, the CPSU's efforts to control and to monitor the project should have ensured that it expressed a coherent worldview, while also streamlining the productive process. In practice, however, it ended up both confusing the actual aims of the project, while also resulting in a major productive gridlock. Trying to preserve ideological orthodoxy meant that the composition of articles and editing almost exclusively went to party members. The problem, though, was that there were only so many experts in both Marxist theory and other fields to write and edit encyclopedia entries efficiently. Those who met these criteria quickly became tremendously overburdened, such that non-Marxist experts still had to be reincorporated into the fold from time to time.[31] The result was that authors would often receive far too many comments, and these comments were frequently contradictory.[32] This resulted in a vicious circle wherein the most rigorous ideological devotion was demanded, but with a lack of clarity as to what the right message actually was, all under the threat of serious repercussions.[33] To try to sidestep these problems, the editors made the rather peculiar decision in 1931 to skip the production of volume 23 all the way to volume 65, which would be the final volume of the first edition; the volumes in between would be released at random until 1938.[34] Volume 65, then, was the first to be released after the purge of the editorial board, though the production of volumes 64 and 63 were plagued by several setbacks, given that they contained crucial entries like those for "Engels," "Ethics," and "Economics." This peculiar dynamic continued throughout much of the course of the existence of the *Soviet Encyclopedia*.[35]

Hessen's own involvement with the *Soviet Encyclopedia* began in 1931, starting with volume 65, suggesting that he was hired onto the project after the mass editorial purge of that year.[36] He entered onto the picture not only as a contributor but also as one of the editors in chief of the physics section,[37] a position which he would hold until his execution in 1936.[38] Further research must be conducted in order to draw a clearer picture of why Hessen, a Deborinite, became involved with the project after the purges, but a few anecdotes nevertheless prove illustrative that his involvement in this endeavor was anything but congenial. For instance, at a meeting of CPSU physicists on March 19, 1931, Vasiliy Egorshin attacked Hessen for, among other things, his *Great* and *Small Soviet Encyclopedias* entries being examples of so-called "mathematical idealism"—a curious, though nevertheless impactful, epithet in the Soviet context.[39] Next, Hessen's article on "Ether" attracted both considerable negative attention and scrutiny. With Einstein's development of special relativity theory in 1905, ether theory was largely seen as superfluous, a position that many younger physicists adopted. Hessen's encyclopedia entry on ether marked an effort to envision a new formulation of the concept beyond its antecedents from the previous centuries. Upon reading Hessen's entry, however, the physicists Matvei Bronstein, George Gamow, Dmitry Ivanenko, Sergei Izmailov, Lev Landau, and I. S. Chumbadze sent a cosigned photo-telegram to Hessen, in which they sarcastically wrote, "[h]aving read your entry on ether started our enthusiastic studies of it. Looking forward to reading about phlogiston," accompanied by an image of a dustbin, containing an old bottle with "phlogiston" written on it and a chamber pot with "ether" written on it.[40] Finally, Hessen's name would appear once more in the *Soviet Encyclopedia* after his execution. Sadly, the very encyclopedia that he was once asked to promote disgraced him as one who "took part in the terrorist, sabotage and espionage work of a fascist Trotskyite-Zinoviev gang,"[41] and who "engaged in advancing Menshevik idealism and idealist bourgeois theories in questions of theoretical natural science."[42]

In the entries themselves, we find a compelling example of how Hessen tried to apply dialectical materialist philosophy to specific scientific concepts and notable figures. Most importantly, they show how new scientific concepts could be understood in a dialectical materialist fashion. His entry on "Space" is primarily an effort to understand how the Einsteinian concept of space-time could be explained through this philosophy, while his entry on "Ether," perhaps his best-known and most controversial entry, functions as a defense of the enduring relevance of this notion in the conflict between quantum mechanics and relativity theory. "Energy" stands out as the lengthiest of Hessen's contributions to the *Soviet Encyclopedia* and is one of the two more substantial in scope, while "Entropy" is the second of Hessen's more comprehensive submissions to the project. "The Ergodic Hypothesis" is the only one of Hessen's contributions to the *Soviet Encyclopedia* that does not incorporate an explicitly dialectical materialist analysis. Save for one throwaway remark about materialism, broadly speaking, the piece is an otherwise rather conventional contribution to the history of science. Hessen's piece on "Einstein" is his only coauthored entry, and it was written with German mathematician, Herman Müntz.[43] Müntz wrote the first part of the article, which comprises the majority of the text's length, and consists of a rather conventional summary of Einstein's biographical details, as well as an account of his contributions to physics.[44] Hessen wrote the second part, which provides a brief synopsis of Einstein's political beliefs as well as a dialectical materialist critique of Einstein's Machist epistemology.[45] While Hessen employs numerous strategies in his entries to the *Soviet Encyclopedia*, I maintain that his conceptual analysis can typically be divided into two parts: first, a critique of a empiricist, Machist epistemological or positivist reading of a scientific concept and second, a dialectical materialist reconstruction of that same concept (or, at least, an outline of how to do so). Throughout them, we again see Hessen's Deborinite argument as outlined previously that newly discovered scientific concepts often initially emerge in an idealistic form, as they initially exceeded the limits of the current materialist paradigm, to then establish the basis of a new understanding of materialism.[46] Thus, they show his understanding of dialectical materialism as one which means perpetually redrawing the parameters of materialism itself. Taken either individually or as a whole, these entries offer compelling insights into the efforts to compose an entirely new system of knowledge in the hands of one of its most gifted thinkers.

NOTES

1. Б. М. Гессен, «Пространство», *Малая Советская Энциклопедия* 6 (1931): 949–51.
2. Б. М. Гессен, «Эфир», *Большая Советская Энциклопедия* 65 (1931): 16–18.
3. Б. М. Гессен, «Энергия», *Большая Советская Энциклопедия* 64 (1933): 427–41.
4. Б. М. Гессен, «Энтропия», *Большая Советская Энциклопедия* 64 (1933): 455–65.
5. Б. М. Гессен, «Эргодическая гипотеза», *Большая Советская Энциклопедия* 64 (1933): 577–78.
6. Б. М. Гессен and Г. М. Мюнтц, «Эйнштейн», *Большая Советская Энциклопедия* 63 (1933): 152–54.

7. Chilvers, "The Tragedy of Comrade Hessen," 113; G. E. Gorelik and V. Ya. Frenkel, *Matvei Petrovich Bronstein and Soviet Theoretical Physics in the Thirties*, trans. V. M. Levina (Basel: Springer Basel AG, 1994), 51; Корсаков, Козенко, and Грачева, *Борис Михайлович Гессен*, 27, 90–91, 169.

8. О. Шмидт, ed., *Большая Советская Энциклопедия* 65 (1931): 3.

9. О. Шмидт, ed., *Большая Советская Энциклопедия* 64 (1933): 4.

10. О. Шмидт, ed., *Большая Советская Энциклопедия* 63 (1933): 4.

11. Н. Л. Мещеряков, ed., *Малая Советская Энциклопедия* 6 (1931): 3

12. Kassof, "A Book of Socialism," 56.

13. P. Josephson, "Boris Hessen as Philosopher and Polemicist: Ideology, Politics and the New Physics in the Soviet Union in the 1920s and 1930s," *Society and Politics – Boris Hessen and the Dialectics of Natural Science* 13.1 (2019): 24.

14. Kassof, "A Book of Socialism," 60; L. Mazliak, "The Beginnings of the *Soviet Encyclopedia*: The Utopia and Misery of Mathematics in the Political Turmoil of the 1920s," *Centaurus* 60 (2018): 32.

15. Kassoff, "A Book of Socialism," 60–61.

16. Ibid., 61.

17. Mazliak, "The Beginnings of the *Soviet Encyclopedia*," 25.

18. F. Weird, "Russia Set to Unveil the World's Newest Print Encyclopedia—And Its Last?," *The Christian Science Monitor*, March 23 2017, http://www.csmonitor.com/.

19. Mazliak, "The Beginnings of the *Soviet Encyclopedia*," 29–30.

20. L. R. Graham, *Science in Russia and the Soviet Union: A Short History* (Cambridge: Cambridge University Press, 1993), 117; Mazliak, "The Beginnings of the *Soviet Encyclopedia*," 25.

21. Mazliak, "The Beginnings of the *Soviet Encyclopedia*," 30.

22. Stalin, "A Year of Great Change," 124–41.

23. Graham, "The Socio-Political Roots of Boris Hessen," 68.

24. Ibid., 70.

25. Ibid., 64, 69.

26. Корсаков, Козенко, and Грачева, *Борис Михайлович Гессен*, 80–81.

27. Kassof, "A Book of Socialism," 68–69.

28. Ibid., 69–70.

29. Ibid., 71.

30. Graham, *Science in Russia and the Soviet Union*, 117.

31. Kassof, "A Book of Socialism," 63, 66–67.

32. Ibid., 74.

33. Ibid., 81.

34. Ibid., 55, 70.

35. Ibid., 75–76.

36. Шмидт, ed., Большая Советская Энциклопедия 65, 5.

37. Gorelik and Frenkel, *Matvei Petrovich Bronstein and Soviet Theoretical Physics in the Thirties*, 51; Chilvers, "The Tragedy of Comrade Hessen," 113.

38. Корсаков, Козенко, and Грачева, *Борис Михайлович Гессен*, 5.

39. Ibid., 90–91.

40. Gorelik and Frenkel, *Matvei Petrovich Bronstein and Soviet Theoretical Physics in the Thirties*, 51; Chilvers, "The Tragedy of Comrade Hessen," 113.

41. Н. Л. Мещеряков, ed., «Меньшевиствующий идеализм», *Малая Советская Энциклопедия* 6 (1937): 822–23. See Корсаков, Козенко, and Грачева, *Борис Михайлович Гессен*, 169.

42. Мещеряков, ed., «Меньшевиствующий идеализм», 822–23. See Корсаков, Козенко, and Грачева, *Борис Михайлович Гессен*, 169.

43. Herman Müntz was a German mathematician, known for his so-called "approximation theorem." Of Jewish ancestry, Müntz fled to the Soviet Union in the years immediately prior to the rise of Nazi Germany and would go on to facilitate the escape of other mathematicians from there thereafter. Müntz would eventually be expelled from the Soviet Union and eventually settled in Sweden, where he lived out the remaining years of his life. E. L. Ortiz and A. Pinkus, "Herman Müntz: A Mathematician's Odyssey," *The Mathematical Intelligencer* 27.1 (2005): 22–31.

44. Гессен and Мюнтц, «Эйнштейн», 152–54.

45. Ibid., 154. In 1919, Einstein signed the founding manifesto of the German Democratic Party, a liberal middle-class party that he never actually joined. Over time, however, Einstein's political views drifted left over time. While initially critical of the Bolshevik Revolution, he would eventually go on to say that "In Lenin I honor a man, who in total sacrifice of his own person has committed his entire energy to realizing social justice. I do not find his methods advisable. One thing is certain, however: men like him are the guardians and renewers of mankind's conscience." Nevertheless, in his latter essay, "Why Socialism?," he appeared to endorse something more akin to democratic socialism than Soviet Marxism. D. E. Rowe and R. Schulman, *Einstein on Politics: His Private Thoughts and Public Stands on Nationalism, Zionism, War, Peace, and the Bomb* (Princeton: Princeton University Press, 2007), 83, 408, 412–13; A. Einstein, "Why Socialism?," *Monthly Review* 7.2 (1955 [1949]): 44–50.

46. See chapter 2.

Space

B. M. Hessen

[949] **SPACE,**[1] along with *time* (see), is the primary form of the existence of matter. Recognizing the existence of matter in motion, independent of our consciousness, materialism recognizes the objective reality of S. and time. The question of S. is expressed in two main trends of philosophy. While materialism recognizes S. as an objective reality, idealism considers it to be a form of consciousness.

For *Kant* (see), S. is an *a priori* (given before experience) form of our sensory perception. The form of S. as the knowing subject's form of perception is imposed upon the sensory world. S. is thereby deprived of objective sense. Kant's ideas are dominant among bourgeois scientists and especially among mathematicians. According to *Mach's* view (see), S. is a way of organizing our sensations ("S[.] and time are ordered systems of sets of sensations"[2]). Taking time and S. for the form of the existence of matter, dialectic materialism recognizes not only the inseparability of S. from time, but time and S. from matter as well. S. and time are filled with matter: empty S. only exists in abstraction. Abstracting S. and time from matter, we turn them into empty forms. Engels elaborates upon the dialectical-materialistic conception of S.: "We know what an hour or a meter is, but not what time and space are. As if time is a sum of hours, and space is a sum of cub. meters. Obviously, the two forms of existence of matter are nothing without matter itself, they are empty abstractions which exist only in our minds"[3] (*Dialectics of Nature*). Motion is a synthesis of S. and time, because motion can only take place in time and S. Recognizing motion as the fundamental and inseparable property of matter, dialectic. materialism establishes not only the inseparability of space from time, but also the inseparability of time, space and matter.

[950] *S. in physics*. Classical physics adopted Newton's concept of S., which he formulated in his main work, *Mathematical Principles of Natural*

Philosophy. The essence of Newton's concept comes down to the fact that he recognizes the objective reality of empty, motionless (absolute) S. It is possible to remove all physical bodies from S. and yet it continues to exist as an empty container for bodies, as a kind of infinitely large box without walls. This is also Newton's perspective on time. We can eliminate all of the processes that unfold within time and yet, time will remain as pure duration. Thus, time and S. exist not only separately from each other, but separately from matter as well. Additionally, Newton admitted the existence of an absolutely motionless S. All of the movements of real things represent relative motion, since they only move in relation to each other. The absolute motion of bodies is motion in an absolutely motionless S. Insofar as Newton recognized the objective reality of S. and time, his theory can be called materialistic. But, since he recognized the independent existence of S. and time, separate and independent from matter, his materialism is metaphysical materialism. Newton's theory was adopted in the mechanical worldview, which prevailed in physics until the beginning of the 20th C., before the appearance of Einstein's *relativity theory* (see). Although the main dialectical conception of S. and time was provided by Hegel at the beginning of the 19th C. and the developed dialectical-materialist. conception was provided by Engels (in the 60s–70s), they did not influence the development of the problem of space-time in physics.

Newton's metaphysical theory only came to be revised with the appearance of relativity theory in physics. The main idea of relativity theory is that absolutely motionless, empty S. does not really exist. S. is inseparable from matter. The movements of individual objects can only occur in relation to each other and not in an absolute S. There is no time as pure, empty duration separate from processes. Also, there is no absolute time flowing independently of processes. Time is inseparable from process as well as S. from matter. And, since physics recognizes and studies only material processes, according relativity theory, time and S. are inseparable from matter. According to relativity theory, the course of processes and the length of material bodies depend on the state of the system in which the processes evolve and in which the bodies are measured. With an increase in the speed of the system, the course of processes slows down and the length of bodies decreases. Thus, the idea of the relativity of time and S. is introduced, since both time and S. depend on the state of the observer. However, this does not mean that relativity theory rests on a subjective point of view, because the synthesis of S. and time in [951] a certain four-dimensional interval turns out to be independent of the state of the observer. Thus, relativity theory not only establishes the inseparability of S. and time from matter, but also establishes the synthesis of S. and time in the concept of a four-dimensional interval. Relativity theory provides a mathematical form for the synthesis of S. and time, which preserves the specific difference of time from S.

While in Newtonian physics the properties of S. and its geometric properties are something absolutely constant and independent of the presence or absence of matter, relativity theory presents the properties of S. as internally related to the configuration of matter. The properties of S. are not absolutely given and independent of matter, but depend on matter and vary based upon changes in the configuration of matter. Bear in mind, however, that by introducing the idea of the synthesis of S. and time from the very outset, relativity theory considers not only the problem of the dependence of space on matter, but also of the dependence of S.-time on matter. In relativity theory, S.-time is a single, four-dimensional continuum. Relativity theory synthetically studies in a mathematical form this four-dimensional continuum, this four-dimensional world. Thus, regarding the basic question of the relationship between S., time and matter, relativity theory generally coincides with the concept of dialectical materialism. It is necessary to separate relativity theory as a purely physical theory from the theoretical and analytical considerations that many physicists associate with it and, partly, Einstein himself. These considerations are similar to Mach's theory, but are by no means a necessary consequence of its physical formulations.

NOTES

1. The following is a translation of Б. М. Гессен, «Пространство», *Малая Советская Энциклопедия* 6 (1931): 949–51. Hessen's original references have been substituted by English translations. This translation was completed with the aid of translation software and revised by O. Bashkina [Translator's Note; hereafter, TN].
2. E. Mach, *The Science of Mechanics: A Critical and Historical Account of Its Development*, 4th ed., trans. T. J. McCormack (Chicago: The Open Court Publishing Co., 1919 [1893]), 506 [TN].
3. Engels, *Dialectics of Nature*, 515 [TN].

Ether

B. M. Hessen

[16] ETHER[1] (phys.), in classical physics, is a special continuous medium in which waves can propagate and which is a carrier of an electric and magnetic field. In early wave-mechanical theories of light (Fresnel), E. was considered [17] an elastic medium that transmitted mechanical vibrations and waves. With the development of the views of Faraday and Maxwell, who denied the possibility of electric and magnetic actions at a distance (long-range), and therefore of action in absolutely empty space, E. became the necessary material carrier of electromagnetic fields. E. differed in its properties from microscopic bodies in that it did not obey the law of gravity, i.e. it did not possess a heavy mass (it was "weightless"). Also, a number of attempts to explain the phenomenon of gravitation by means of motion and distortions in E. have not yet yielded any results.

The development of the electrodynamics of moving bodies led classical physics to view E. as a medium that is not carried away by any bodies when they move within it. This is how the Lorentz theorem was created, according to which E. fills the entire universe as a whole, and as a whole, is absolutely motionless; only electromagnetic disturbances (waves) can occur within it. Such a view of E. made it possible to determine the absolute motion of bodies, because if E. is in a state of absolute rest, then all motion relative to it is absolute motion, and rest is absolute rest. Therefore, if it were possible to determine the speed of the earth in relation to E., this would be the speed of its absolute motion, and all other speeds (i.e., movements in relation to various celestial bodies) would be relative speeds and movements.

Michelson's experiments (see), which set out to determine the absolute speed of the earth in relation to E., did not yield any results. Continued repetitions of these experiments (Morley, Miller) were also inconclusive. All of this, as well as a number of problems in the electrodynamics of moving bodies, led to the emergence of Einstein's special theory of relativity. Since Michelson's experiments do not allow for the possibility of determining the motion of bodies relative to E., special relativity theory declares E. to be non-existent; and since at the same time, it does not return to action at a distance, but upholds the point of view of short-range action, in order to overcome this difficulty, special relativity theory resorts to pure mathematical description, refusing to consider the issue of the medium in which electromagnetic phenomena occur, thereby refusing to answer the question of the objectivity of physical phenomena, i.e. essentially adopting Mach's point of view on the question of E.

The opposite view, advocated by the school of J.J. Thomson, is essentially a return to the views of Faraday and seeks to interpret electromagnetic phenomena mechanically. Mechanical theories recognize the objective reality of E., but in turn, lead to irresolvable contradictions. So, i.e. it follows as an inevitable consequence from W. Thomson's theory that E. must be fixed at its limits—otherwise it will all gather at one point similar to unfixed soap [18] foam. At the same time, it is completely incomprehensible and unthinkable how, given the infinity of the universe in space, E. can be fixed at its limits.

The problem of E. is one of the most difficult problems in physics and although general relativity theory was forced to once again restore ether as a carrier of material fields, modern physics still continues to be dominated by purely phenomenological views on ether, which essentially—in one form or another—deny its existence.

The primary methodological error of general relativity theory is that it considers E. to be an absolutely continuous medium, while all modern theories of the structure of matter (quantum theory, wave mechanics) more and more confirm the position of dialectical materialism, according to which matter in its objective structure is simultaneously continuous and discontinuous.

The absence of the idea of discontinuity in general relativity theory and the metaphysical absolutization of continuity close the path for further development of the study of E. New theories of the structure of matter, too, only outline the path to solving the problem of E., because in general they cannot yet produce a synthesis of electromagnetic waves and photons (quantum of light) on one hand, and electrons and waves of matter on the other (on this, see the articles *Electron* and *Electromagnetic Theory of Light*). In their present state, these theories are not devoid of a significant amount of mathematical formalism. Thus, the problem of ether in modern physics has only been posed, and is by no means resolved—even in a general way.

In physics, we often encounter a completely incorrect opposition of E. to matter. In *Dialectics of Nature*, Engels says of E. that "[i]f it *exists* at all, it must be of a material nature, it must come under the concept of matter."[2]

Taking the heavy and inert mass as the only criteria of materiality, physicists were inclined to deny the materiality of E., because it has neither heavy nor inert mass. Here, we have confusion between the physical and philosophical concepts of matter, an exhaustive analysis of which was provided by Lenin in his consideration of the crisis of physics at the beginning of the 20th Century (see *The Crisis of Natural Science*). E. is a qualitatively special type of matter, and physics only now starts approaching the study of its intrinsic properties. But, it has the same objective reality as all other material bodies. From this point of view, the opposition of E. to matter is senseless and leads to agnostic and idealistic conclusions. See *Relativity Theory*, *Electromagnetic Theory of Light*.

NOTES

1. The following is a translation of Б. М. Гессен, «Эфир», *Большая Советская Энциклопедия* 65 (1931): 16–18. Hessen's original references have been substituted by English translations. This translation was completed with the aid of translation software and revised by O. Bashkina [Translator's Note; hereafter, TN].
2. Engels, *Dialectics of Nature*, 523 [TN].

Energy

B. M. Hessen

[427] GENERAL DEFINITION OF E.

E.[1] (from the Greek word *energeia* – activity, capacity for work) one of the principal phys. categories. The *law of energy conservation and transformation* (see) is of fundamental importance for establishing the concept of E. and for determining its numerical value. In physics, E. is usually defined as the ability to accomplish work. This definition does not establish the connection between E. and the transformation of one form of motion into another. However, it does essentially presuppose the need to transform [428] one form of motion into another. Indeed, work can only be accomplished by the disappearance of some form of motion and its transformation into another. If a body moves by inertia without friction, it retains its motion (momentum) and kinetic E., but it does not accomplish any work. Work is only accomplished through the reduction of kinetic E., when (in the absence of a force field) the forces of friction appear. Engels, who defines energy as the measure of motion during its transformation from one form into another, very clearly established and developed the connection between different categories of E. with the problem of the transformation of one form into another (see §7).

THE RELATIONSHIP BETWEEN E. AND THE LAW OF E. CONSERVATION. TYPES OF E. IN CLASSICAL PHYSICS

The concept of E. can only be understood on the basis of the law of E. conservation and transformation. If we observe a decrease in some kind of E., e.g. the kinetic energy of a body in motion, we conclude on the basis of this law that this decrease causes the appearance of a different kind. If a body with a certain amount of kinetic E. moves upward in a gravitational field, its kinetic E. decreases over time and eventually becomes zero. Based on the law of E. conservation, we consider this process to be the transformation of kinetic E. into another form, e.g. potential E., which, in contrast to kinetic E., is characterized by the velocity of the motion of a body, and by its position in the gravitational field.

If a decrease in kinetic E. occurs due to friction against motion, a decrease in kinetic E. marks its transformation into a new form, e.g. thermal E., that, considered macroscopically, cannot be characterized by either velocity or position, but by a new quantity, e.g. temperature. If the appearance of thermal E. from friction does not account for the entire balance of E., then a new type of E. must be introduced, e.g. electrical E., such as when static electricity appears, which in turn bears a new characteristic, e.g. electrical charge. In doing so, classical physics considered mechanical, thermal, electrical and magnetic energy to be primary types.

Deferring a detailed exposition and derivation of the quantitative expressions of the various forms of E. to their respective articles (see *Mechanics*, *Thermodynamics*, *Electricity*), we instead focus here on the relationship between mechanical E. and other types, which was of great importance in the history of physics.

"Motion, as applied to matter, is change in general."[2] Any change that occurs in matter can be characterized as a transformation from one state into another. Various forms of the motion of matter can be considered changes of state. Therefore, in order to study the quantitative aspect of motion, it is necessary to be able to determine those characteristic quantities (parameters) that define this form of motion. Then, the change of state will be characterized as a change in these parameters, and will be quantified through the parameters that determine that state. In those cases, [429] when several processes take place simultaneously in a material body (e.g. the body moves in space and heats up), which occurs in all real cases, the general E. of all processes is additive. The additivity of various forms of E. presupposes the independence of various ways of affecting a body, e.g. we can consider the heating of a moving body independent of its mechanical motion and the effect on its mechanical motion independent of heating. In this sense, the so-called principle of the superposition of E. (Planck's term) suggests that any combination of influences on a body can be decomposed into independent components that separately change the components of a body's overall E. In general, the superposition principle is, strictly speaking, incorrect since it is not always possible to decompose the effect into a number of different components of E. But in ordinary cases, with a sufficient degree of approximation, the superposition principle can be considered correct.

The very term "energy" was first introduced into science by Young as a substitute for the term "living force" (*vis viva*), which had been used since the time of Leibniz. Young only uses the term E. in the sense of kinetic E., where E. denotes "the ability to produce work due to acquired velocity." The term "potential" E., according to Tet, was first used by Lazare Carnot, who introduced the difference between kinetic and potential E., referring to the latter as "latent living force." The terms were introduced into general use by Rankine. Rankine defines kinetic E. as "active" and potential as "stored energy." This definition shows a close connection between the definition of potential E. and conservative systems.

For *conservative systems* (see) with fixed Cartesian coordinates, kinetic E. is defined as a pure quadratic function of velocities $T = \Sigma a_{ik} v_i v_k$ and potential as a pure quadratic function of coordinates $\boldsymbol{P = \Phi (x, y, z)}$. The quantities that determine mechanical motion are position (coordinates) and velocity; since potential and kinetic E. are determined by the same properties, these types of E. are usually attributed to mechanical E. In the case of fixed Cartesian coordinates, we have the following expression for E. (for a material point). Kinetic E.:

$$L = \frac{1}{2} m \left(x^2 + y^2 + z^2\right).$$

For fixed Cartesian coordinates, $\boldsymbol{L}$ is a homogenous function of the squares of the velocities. For moving Cartesian coordinates, the expression $\boldsymbol{L}$ can include terms which contain the velocities to the first power, or terms that do not contain velocities at all. For example, for rotating coordinates (where the $\boldsymbol{Z}$-axis is fixed), we have the following expression for $\boldsymbol{L}$:

$$L = \frac{m}{2} \left(x_1^2 + y_1^2 + z_1^2\right) + 2a \left(x_1 y_1 - x_1 y_1\right) + a^2 \left(x_1^2 + y_1^2\right),$$

where $\boldsymbol{a}$ refers to the angular velocity of rotation of the coordinate axes. Thus, for mechanical motion, kinetic E. is solely a function of velocity only in the case of fixed Cartesian coordinates.

In the presence of a force field—expressed by the forces $\boldsymbol{F (x, y, z)}$ acting on a material body, which are functions of the coordinates of a point—, potential is added to kinetic E. If $\boldsymbol{X, Y, Z}$ are forces acting on a body, and $\boldsymbol{dx}$, $\boldsymbol{dy}$, $\boldsymbol{dz}$ are changes in position under these forces, then the potential E., in of the central forces, can be expressed as:

$$-d\Phi = \lambda dx + Ydy + Zdz.$$

For central forces and conservative systems, the expression $Xdx + Ydy + Zdz$ will be the total differential, and Φ will be a pure function of coordinates:

$$\Phi = -\int Xdx + Ydy + Zdz = \Phi\,(x, y, z).$$

For a gravitational field (for a material point with one degree of freedom), $Z = -mg$ and $\Phi = mgh$. The total energy of mechanical [430] motion E is equal to the sum of kinetic and potential: $E = L + \Phi$.

From both expressions (1) and (2), an expression can be derived for potential and kinetic, as well as the total E. for all specific cases of mechanical motion, provided that the equations of motion are given as one of the integrals of these equations. For example, in the case of simple oscillatory motion (simple harmonic oscillator), $mx + k^2x = 0$ (where k^2x is the elastic force) and

$$E = \frac{m}{2}x^2 + \frac{m\,(2\pi v)^2 x^2}{2},$$

where x is the displacement from the equilibrium position, and v is the oscillation frequency:

$$(2\pi v)^2 = \frac{k^2}{m}$$

(for more details, see *Elasticity Theory*, *Dynamics*).

The total E. of a conservative system can always be decomposed into potential and kinetic. Since all isolated mechanical systems are conservative, their common E. can always be represented as the sum of potential and kinetic, with both having a certain physical value. For non-conservative systems (e.g. in the case of a system where the acting forces are not purely functions of distance, but also dependent on velocities, such as, e.g. forces of friction), the division of total E. into kinetic and potential becomes more or less contingent. In the question of E., the mechanistic worldview sought to reduce all types of E. to potential and kinetic in accordance with the desire to reduce the entire variety of forms of motion to mechanical motion. Therefore, e.g. magnetic E. was considered to be kinetic, and electrical to be potential. However, this distinction was only based on an external analogy with mechanical systems. When the Lagrange equations is used for electromagnetic phenomena, the Lagrange function can be represented in some cases as the difference between two quantities; one being a quadratic function of generalized velocities, and the other a function of generalized coordinates. On this basis, by analogy with the Lagrange function for mechanic systems, one kind of E. is taken for kinetic and another for potential. The conventionality of this reasoning is evident from the fact that when considering electric. processes, it is valid with constant self-induction. In the case of variable self-induction, magnetic E. will enter the equations of motion as potential.

The convention of dividing any E. into potential and kinetic can also be seen from the fact that in a number of cases, when applying the principle of least action, it is necessary to consider the expression of kinetic and potential as a whole, without dividing it into potential and kinetic E. In the case of magnetic E., in order to interpret it as kinetic, we have to resort to the hypothesis of the motion of "hidden masses," with the character and nature of these masses and their motion being essentially unknown. In the history of physics, there have also been a number of unsuccessful attempts to reduce potential E. to the kinetic E. of free particles in the absence of a force field. All such attempts are based on a purely mechanical model of the force field, such as e.g. is found in Le Sage's mechanical theory of *gravitation* (see), according to which gravity is the result of the impacts of particles flying from outer space. Potential E. is associated with the presence of a force field, i.e. forces of interaction, which are functions of a particle. The problem of the elimination of potential E. is therefore connected with the problem of the elimination of force.

The elimination [431] of forces of interaction and the reduction of potential E. to kinetic is not always possible in as simple of a way as is seen in Le Sage's example. General relativity theory attempts to eliminate gravitational forces and, therefore, to provide a different representation of the force of the gravitational field. Therefore, in general relativity theory, in a relatively small area of the gravitational field, it is possible to eliminate potential E. Heat, regarded as the mechanical E. of molecules and atoms, appears in the form of kinetic E. in the case of an ideal gas and as a combination of kinetic and potential E. in the case of a solid body. However, it would be wrong to see a simple reduction of thermal into mechanical energy in the kinetic theory of heat. Thermal motion is characterized not only by the mechanical motion of elementary particles, but also by the disorderly (chaotic) nature of these movements. It is the chaotic nature of thermal motion (molecular chaos) that distinguishes the thermal form of motion from the organized mechanical motion of a system of particles and is the basis for the specific patterns in thermal phenomena.

In the reasoning of thermodynamics, E. is defined according to macroscopic, purely phenomenological parameters. Heat is not seen as a form of motion. Considering heat as a form of the motion of matter leads us to the kinetic theory of heat, which reveals the meaning of the macroscopic parameters. The temperature in the simplest case of an ideal gas acts as an expression of the kinetic E. of the disordered motion of molecules. The internal E. of a gas acts as the sum of the kinetic E. of molecules. Internal E. can be expressed in the following way (for a stationary state):

$$E = \frac{m}{2} \sum n_i c_i^2 = \frac{m}{2} Nc^2,$$

where m is the mass of a molecule, n_i is the number of molecules within a given range of velocities (in a stationary state), and c is the average velocity of molecules in a stationary state (see *Thermodynamics, Kinetic Theory, Statistical Mechanics*).

E. AND EQUATIONS OF MOTION

Despite the tremendously important role that E. plays in all physical processes, it would be wrong to regard it as a universal physical category, from which one can obtain all other categories and equations of physics. Such attempts have been carried out in the history of physics. An entire trend of energy theorists tried to derive the equations of motion of a general type and a number of theses in various fields of physics from the "energy principle," or the law of E. conservation. The best-known work along these lines was Helm's. Boltzmann criticized Helm's work. He showed that Helm's conclusions are not only based on a mathematical error, but also implicitly introduced a number of new assumptions.

Helm's conclusion is as follows. He assumes that kinetic E. is given by the typical expression

$$L = \frac{1}{2} m (x^2 + y^2 + z^2),$$

and the differential of work by $dA = Xdx + Ydy + Zdz$; thus, $dL = m (xdx + ydy + zdz)$. Equating both differentials, we get:

$$(mx - X)\, dx + (my - Y)\, dy + (mz - Z)\, dz = 0.$$

Then considering, dx, dy, dz as independent quantities and establishing that $dy = dz = 0$, Helm obtains $mx - X = 0$, i.e., the standard equation of motion; then equating $dx = dz = 0$, he derives the equation of motion for the Y coordinate, etc. The fallacy of this conclusion is in the fact that, assuming dx, dy, dz to be independent, but at the same time assuming $dy = dz = 0$, he assumes that $y = z = 0$, since when $dx = xdt$, $dy = ydt$, $dz = zdt$. In other words, Helm confuses variation (δ) with the differential (d).

[432] Therefore, it turns out that he does not obtain the equation of motion, but only the law of motion in the direction x, under the assumption that the coordinates of the velocities along y and z are equal to zero. Thus, Helm's equation for E. does not allow us to obtain the equations for a general case

of motion, but only the equations of motion for a one-dimensional case (one degree of freedom).

The impossibility of obtaining the equations of motion from the law of E. conservation is clear from the following simple considerations. Let's assume that a material body moves freely. According to the law of energy conservation, its velocity must remain constant (since kinetic E. remains constant). However, from the constancy of velocity, we cannot draw any conclusion about the actual trajectory of motion. Indeed, the trajectory can be either circular or rectilinear. In both cases, the velocity will be constant. Therefore, we cannot judge the direction of the velocity based on the magnitude and constancy of kinetic E., since kinetic E. as a scalar does not depend on direction. This explains why Helm could only actually obtain the equation of motion for a one-dimensional case, i.e. for the case when there is no fundamental difference between a scalar and a vector. The equations of motion for reversible processes, like the law of E. conservation in conservative systems, can be deduced from the principle of least action. This explains the great significance of this principle in its various forms in physics (see *Variational Methods*).

THE INERTIA OF E. THE RELATIONSHIP OF E. AND MASS

Classical physics rested on the standpoint of the absolute separation of mass and E. It recognized the possibility of the existence of mass without E. (e.g. atoms at absolute zero) and the possibility of the existence of E. without mass, e.g. electromagnetic radiation. According to the views of classical physics, if radiation with E. E bombarded and was completely absorbed by a plate of a certain mass, possessing a certain reserve of thermal E, then the E. of the plate would increase by the value of E, but the mass of the plate would not change as a result. The same was the case with mechanical motion: a body of mass $\boldsymbol{m}$, moving with a velocity $\boldsymbol{v}$ (in the absence of a force field) has an E. equal to $\frac{1}{2}\mathbf{mv^2}$. If this body possesses a high velocity, then its E. will increase, but the mass will remain unchanged. Modern physics considers the relationship between mass and E. in a different way. It accepts that the mass of a body depends on its amount of E., i.e. that E. has inertial mass. Thus, imparting E. to a body, e.g. by heating a body, we increase its mass. This is the postulate of the inertia of E., which is of fundamental importance to all of modern physics.

One of the most important consequences of the inertia of E. is so-called mass defect. If, as modern views hold, all elements are composed of electrons and protons, and if complex elements are composed of simpler ones, then the deviation of the values of atomic weights from whole numbers would be inexplicable. If e.g. an oxygen atom is formed from 16 hydrogen atoms or 4 helium atoms, then to explain why the atomic weight of oxygen (with respect to H = 1) is not 16 but 15.87, it would be sufficient to assume that during the synthesis of oxygen from hydrogen atoms, a radiation of E. is released, the mass of which is equal to the mass defect of oxygen.

[433] The numerical ratio between mass and E. is expressed by the formula $m = E/c^2$ (*c* – the velocity of light in a vacuum).

Mass is one of the most common physical properties of all forms of the motion with which we deal in physics. Whatever form of motion matter possesses, we always find that it has mass (classical physics made an exception for electromagnetic E., but we now know that radiation also has mass). However, this is by no means a return to the old concept of "weightless matter." If, as they say, "matter passes into radiation," this does not at all mean that mass is altogether annihilated and that radiation appears instead. Radiation also has mass, such that in the transformation of mass into radiation, we do not have the disappearance of mass, but the transformation of one form of motion into another.

The sun continuously radiates E. As a result, its mass decreases. But the mass of the emitted E. is exactly equal to the mass lost by the sun. The difference is that a complex set of forms of motion, which is the sun, is converted into, among other things, electromagnetic radiation which has, along with other physical characteristics, the most common property of mass. Therefore, clearly, if we want to establish a quantitative relationship between mass as a characteristic of radiation and the change in the mass of the sun, we will have to resort to the ratio $m = E/c^2$. Knowing the amount of emitted E., we have the opportunity to determine how much the mass of the radiation source has decreased due to radiation. But at the same time, due to the law of mass conservation, this expression allows us to determine the inertial mass of radiation. Establishing the inertia of E. therefore makes it possible to connect the law of mass conservation (sometimes inaccurately called the 'law of matter conservation') and the law of E. conservation (see *Relativity Theory*).

THE DISCONTINUITY OF E. AND QUANTUM THEORY

One often encounters the assertion that quantum theory establishes the atomicity of E. as opposed to the views of classical physics, which considers E. to be continuous, and that in this respect that quantum theory establishes the same atomic structure of E. as we have with respect to matter in atoms and molecules. Undoubtedly, in Bohr's first interpretation of quantum theory, which is limited to conditionally periodic systems, introduces discontinuity into classical physics, which it establishes along with the continuity of all processes, while classical physics only recognized pure continuity. However, it does not at all follow from this that quantum theory establishes the bare atomism of E. First of all, note that the primary discrete quantity that enters into all quantum laws is not E., but action. The quantity $\boldsymbol{h}$, the universal constant first introduced by Planck, is not a quantum of E., but a quantum of action. This is of fundamental importance for correctly understanding the problem of the continuity and discontinuity [434] of E. The quantity of "action" represents the product of the value of E. by the time interval $\boldsymbol{Et}$. Thus, the period of time included in the expression of the action is that period of time during which the motion took place (if we mean simple mechanical motion), or this interaction, if we have a process of transformation from one form of motion into another. Also note that, unlike E., the action is invariant with respect to the Galileo-Lorentz transformation group.

The constant $\boldsymbol{h}$ introduced by Planck is a quantum of action, not a quantum of E., and represents the product of E. by a period of time. Therefore, the atomicity of this quantity does not at all follow from the atomicity of E. In fact, we can assume that both $\boldsymbol{E}$ and $\boldsymbol{t}$ vary continuously, but that their mutual dependence is such that the product of their numerical values always (i.e., whatever the nature of the interaction) gives a constant value. One of the cardinal differences between quantum theory and classical physics is that quantum theory connects E. with frequency by the known correlation between the quantum E. through the quantum of action and frequency ($\boldsymbol{\nu}$):

$$\boldsymbol{E = h\nu.}$$

This is fundamental to quantum mechanics. The main difference between quantum mechanics and classical quantum theory regarding E. is that in quantum mechanics, the accuracy of the measurement of E. depends on the time interval in which the measurement is made. The more precisely we want to determine the value of E., the longer the period of time we need; conversely, the shorter the time interval, the larger the number of values of E. must be taken. In the limit case, in an infinitely small time interval, only infinitely large values of E. can be measured.

In classical physics, the value of E. can be measured with any accuracy at any time interval, including an infinitely small one. A more detailed consideration of this issue leads us to the so-called "uncertainty principle" of *quantum mechanics* (see). From the discontinuity of the quantum of action, it is by no means possible to conclude the absolute discreteness of E. and proceed, as is sometimes done, from analogy with the atomic structure of matter, by simply transferring all of these concepts to E. When we talk about "quanta of E."

in quantum theory, the situation is different than with the quantum of action. According to quantum theory, (with the restrictions mentioned above) the source cannot emit just any amount of E., but only that which is equal to the product of the frequency and ***h***: $\boldsymbol{E = h\nu}$ (***h*** is Planck's constant—the quantum of action). Thus (from the point of view of Bohr's first theory), if we are given a source of radiation, this particular source can only emit E. in discontinuous portions. But, it still does not follow from this that every possible E. can be broken down into such portions (quanta). Indeed, since the quantum of E. is the product of ***h*** by the frequency, and since frequency is a quantity that can change continuously, then from whatever quantity of E. we are given, we can always choose a source that would emit exactly this portion of E.

The essence of quantum theory, therefore, lay not at all in the fact that it introduces the concept of the discontinuity of E. in general, but that in contrast [435] to the views of classical physics, according to which any source could in principle radiate E. continuously—more precisely, the power of its E. could change due to continuous radiation—, according to quantum theory, a given source can only emit E. in certain portions (quanta); that is, it can only be in a number of discrete energy states.

The one-sided emphasis on the discontinuity of energetic processes and the mechanical transfer of the concept of the discontinuity of E. generally leads to an incorrect approach to the problem of E. and matter, since they overlook the main problem of the synthesis of continuity and discontinuity. The entire development of modern physics shows the limited point of view of reducing all physical phenomena to either pure discontinuity or pure continuity. On the contrary, the correct interpretation of the latest changes taking place in the theory of the structure of matter (quantum mechanics, quantum electrodynamics) confirms the point of view of dialectical materialism regarding the synthesis of opposites; that is, of discontinuity and continuity (see *Quantum Mechanics*).

ON THE HISTORY OF THE QUESTION

Various views on E. can be found in the history of physics. The common definition of E. has been closely connected with the fundamental principles of Newtonian mechanics and with the role of force in Newton's system. This definition of E. became widespread in physics, due in large part to Helmholtz's work on the conservation of E. In contrast to Helmholtz's conception, R. Mayer associates the problem of E. with forms of motion, and not with force. The concept of E. in physics was brought to the fore in the era when—thanks esp. to their technical applications—the basic forms of motion had already been studied in physics: mechanical, thermal and electromagnetic. The concept of E. was inextricably linked with the development of the law of E. conservation. Some scientists wanted to see this law as a simple statement of equivalents where one form of E. transforms into another. The law of E. conservation, and with it the concept of E., was not only not considered one of the main methodological prerequisites of natural science, but essays trying to interpret this law as a fundamental law of nature, inextricably linked with the general worldview, were repressed. The antithesis of Helmholtz's and Mayer's concepts is not an accidental historical fact. The beginning of this opposition can be traced throughout the history of physics, starting with Newton. Helmholtz's general methodological concept differs little in essence from Newton's concept. In addition to reducing all forms of motion to mechanical, we find in Newton the separation of motion from matter and the transfer of the motive principle outward, to the applied forces. Leibniz takes the opposite position on this issue. Arguing against a purely mechanical explanation of nature, Leibniz writes to Sturm that he sees the most important confirmation of his views in that "the *foundation of the laws of nature* . . . is not to be found in the principle that the same quantity of motion is conserved . . ., but rather in the principle that the *same quantity of active* [436] *power is necessarily conserved*, and indeed—a thing which I discovered happens for a most beautiful reason—the *same quantity of motive action* also, the value of which is far different from what the Cartesians think of as the quantity of motion."[3] For Leibniz, force is not something external, outside of matter. Lenin emphasizes that, in contrast to Spinoza, "in Leibnitz there is, in addition to the concept of substance, the *c o n c e p t o f f o r c e* 'and indeed of active force, [therefore]"[4] Lenin notes, "Leibnitz through theology arrived at the principle of the inseparable (and universal, absolute) connection of matter and motion."[5]

Lenin further points out the passage from Feuerbach's history of philosophy, where Feuerbach says that unlike Descartes, for Leibniz, bodily substance is no longer an extended, dead mass, set in motion from the outside, but as a substance, bears an active force that does not know rest from its activities. At this point, Lenin makes an important remark: "[f]or this, to be sure, Marx valued Leibnitz, despite his, Leibnitz's, 'Lassallean' features and his conciliatory tendencies in politics and religion."[6]

This difference in the concepts of the relationship between matter and motion is revealed both in the difference of interpretation of E. and in the meaning that this physical category plays in Leibniz and Newton. Although the roots of the concept of E. are found in Stevin and Galileo, E. does not play any role in Newtonian mechanics. This is primarily due to the fact that all of Newton's mechanics mainly interprets the laws of purely mechanical motion, and does not pose the problem of transforming the mechanical form of E. into other forms. This interpretation of Newton can be explained by the fact that the socio-economic formation of his era posed purely mechanical problems as the main tasks of physics. As soon as the steam engine began to be developed after the industrial revolution, and with it thermodynamics, the problem of transforming the thermal form of motion into the mechanical one came to fruition (see *Entropy*). The development of this problem was stimulated by the gradual highlighting of the electrical and chemical forms of motion. This led to the final formulation of the "law of E. conservation." However, two lines can be distinguished with respect to the interpretation of the concept of E. in the history of physics. One proceeds from Leibniz and Huygens to Robert Mayer and highlights the notion of "vital force" or kinetic energy, the other from Newton to Helmholtz and highlights the concept of force in the Newtonian sense. Philosophical theories have played a prominent role in the development and opposition of these two tendencies.

Helmholtz's methodological approach to the problem of the law of E. conservation is determined entirely by his mechanical worldview, which is directly descended from Newton's point of view. "The existence of matter," Helmholtz writes in an introduction, "in itself is to us something tranquil and devoid of action."[7] Thus, in Helmholtz, motion appears external to matter. "[T]he problem before us is to refer back the phænomena of nature," [437] Helmholtz continues, "to unchangeable final causes. This requirement may now be expressed by saying that for final causes unchangeable forces must be found."[8] Thus, force is not only external to matter, but also something absolutely unchanging. Helmholtz further interprets motion as mechanical motion: "[m]otion is the alteration of the conditions of space."[9] By separating motion from matter in this way and by absolutizing forces, Helmholtz sees the task of natural science as that "[t]he phænomena of nature are to be referred back to motions of material particles possessing unchangeable moving forces, which are dependent upon conditions of space alone."[10] This formulation is literally no different from Newton's. As Engels notes, in this formulation of the question, "there is to be found nothing new for the position in 1847 beyond the . . ., mathematically very valuable, development that 'conservation of force' and central action of the forces active between the various bodies of a system are only two different expressions for the same thing."[11]

R. Mayer's point of view advances the problem of the transformation of one form of motion into another as a universal law of nature. "To become acquainted with force in its different *forms*, to investigate the conditions of its *metamorphoses*—this is the only task of physics."[12] "It can be proven *a priori* and confirmed everywhere by experience that the different forces can be *transformed* into one another."[13] Everywhere, R. Mayer uses the concept of force in the Leibnizian sense of "vital forces." It is enough to compare Mayer's and Helmholtz's formulations to see the superiority of Mayer's points of view, which clearly poses the problem of the universal connection of the transformations of forms of motion as opposed Helmholtz's reduction to mechanical movements (see *Law of Energy Conservation and Transformation*). Mayer extends the law of E. transformation to all forms of motion, including the organic. He also considers cosmic processes from this point of view, correctly noting the decelerating effect of tidal friction on the rotation of the earth, and for the first time explained the incandescence of meteors when they fall to the Earth's atmosphere by the transition of mechanical E. to thermal (friction against air).

The struggle waged around the law of E. conservation at the time of its discovery is still ongoing. In his work on the law of E. conservation written in 1887, M. Planck draws special attention to the fact that Mayer was able to understand that Leibniz's concept of force has the advantage over Newton's, since it was only from this point of view that it was possible to establish the relationship between matter and motion, between matter and E. Furthermore, Planck indicates that due to the fact that all physics at that time was based on mechanics, Newton's concept of force was universally recognized, and Mayer's ideas were forgotten.

In his work, *Mechanics*, which appeared earlier than Planck's, Mach approaches the problem of force and E., and evaluates R. Mayer in a completely different way. According to Mach, the main reason that force was afforded predominant importance in physics, rather than E. (work), was [438] Galileo's mistake when, in deriving the laws of freefalling bodies, he suggested that total velocity is proportional to distance traveled, and not to the square root of distance traveled, i.e.

$$ds/dt = as, \text{ but not } ds/dt = a\sqrt{s}.$$

"It was only gradually and with great difficulty," Mach writes, "that the concept of 'work' attained its present position of importance; and in our judgment it is to [this] trifling historical circumstance that the difficulties and obstacles it had to encounter are to be ascribed."[14]

The essence of the matter, according to Mach, is that the equation $v = gt$ was discovered earlier and therefore, the equation $s = gt^2/2$ was considered its immediate consequence, and $gs = v^2/2$ the more distant. With the introduction of mass and force, three basic equations of mechanics were obtained: (1) $mv = pt$; (2) $ms = pt^2/2$; $ps = mv^2/2$, and since equations (1) and (2) were discovered earlier, the concepts of force and momentum were considered fundamental. This explains, according to Mach, the dispute between Leibniz and Descartes about the two measures of motion: mv and mv^2.

So, Mach is not only incapable of posing the problem of the interrelation of forms of motion, but he reduces the entire principle aspect of the dispute about force, energy and motion to an accidental error by Galileo. Evaluating the work of Robert Mayer, Mach, although giving him credit for the fact that he prioritizes E. and not Newtonian force, in accordance with his general concept, summarizes his opinion on Mayer by emphasizing that "Mayer's error was, that he regarded his method as the only correct one."[15]

ENGELS' CONCEPT OF E. AS A MEASURE OF MOTION

It is very important to clarify the difference between Marx and Engels' understanding of measure and the ordinary usage of the term in physics. Usually, they pay attention to the quantitative side of measure and imagine it to be an abstract quantity. Such an interpretation omits the essential aspect that the problem of measure arises from the problem of the interrelationship between two mutually connected, but qualitatively different quantities. Therefore, measure can be understood if we consider it not just from the point of view of its quantitative expression, but also from the point of view of its qualitative determination, i.e., we clearly imagine that a measure is a "qualitative quantity" that arises in the process of the interrelation and connection between qualitatively different and at the same time, quantitatively comparable objects. See *Measurement* for details.

When posing the problem, the relationship between two qualitatively different forms of the motion of matter is equally important as the fact that they are mutually connected, since they are all based on matter in motion and the fact that they are qualitatively different, i.e. are irreducible to each other. The very problem of the relationship is, of course, essentially determined by the fact that we have a qualitative variety of forms of motion in nature, but the solution to the problem of the relationship is only possible because they all form a unity through matter in motion. This [439] formulation of the question excludes not only the interpretation of the unity of the forms of motion, as the reduction of all forms of motion to mechanical motion, as found in the mechanical worldview and in the mechanistic interpretation of the law of E. conservation, but also the view that various forms of motion are completely separate entities, which was commonplace prior to the discovery of the law of E. conservation. Thus, the problem of the measure of various forms of motion is based on the law of universal connection and transformation of one form of the motion of matter into another, since only by ascending to the understanding of the unity of qualitatively diverse forms of motion of matter can the question of the measure of motion be properly raised. Therefore, the expression of various forms of E. can only be obtained on the basis of the law of E. conservation.

That is why Engels connects his conception with the law of E. conservation and transformation and at the same time points out that the indestructibility of motion and E. must be understood not only in a quantitative, but also in a qualitative sense. Indeed, only such an understanding leads us to the correct formulation of the problem of measure (see *Entropy*, §6). Considered in this respect, E. is an expression of the unity of all forms of motion, though not in the sense of their qualitative homogeneity, but in the sense of their mutual transformations into each other.

Therefore, E. does not characterize any particular type of motion, but the general connection between the forms of motion, in their mutual transformations. This expresses the qualitative aspect of E. as a measure of motion. But, on the other hand, E. must appear in its quantitative determination; in other words, it must appear as a definite form of motion, quantitatively measured according to its inherent characteristics. The fact that E. in its quantitative determination should appear with the characteristics of a particular form of motion does not in the least contradict its universal significance as a measure of motion across all transformations.

E. is not a special substance; it is a measure and therefore expresses a ratio. But, the quantitative expression of this ratio is always realized in some specific form of motion. When we say that a body of mass ***m***, moving with a velocity ***v***, has an E. equal to $\frac{1}{2}mv^2$, this means that we can indicate the amount of another form of motion into which this mechanical motion can pass. Indeed, a moving body can only do work if its mechanical motion is eliminated, since it will transform into another form of motion, e.g. heat.

Never and nowhere can there be a[n] appearance of E. and work if there is no transformation of one form of motion into another. But, the amount of work is the quantitative expression of E. So, the expression of E. according to the properties of any particular form of motion (e.g. the mechanical equivalent) only makes sense because we can unify two qualitatively different forms of motion by an internal connection based on the possibility of mutual transformations. If we say that the mechanical equivalent of heat is 427 *kgm*,[16] then by this we indicate that the amount of the mechanical form of motion, measured by the inherent properties of this form [440] of motion (mass, velocity, position in space) into which a certain amount of the thermal form of motion will turn, measured by the inherent properties of this form of motion (amount of heat, expressed in heat capacity and temperature). Therefore, we can only accomplish this, e.g. the mechanical equivalent of the general quantitative expression of E., by virtue of the universal connection of the forms of motion.

Thus, the quantitative expression of E. can be given according to the properties of any form of motion, but always bear in mind that these are not different "energies as substances," but different numerical expressions of qualitatively different forms of motion that are universally connected and can mutually transform, one into the other. Therefore, the expression "matter possesses E." must be understood not only in the sense that E. can be expressed according to properties which are specific to a given form of motion. The primary meaning of this expression is that possessing E. means the possibility of transforming a given form of motion into another. And, it is precisely this position that forms the basis for understanding E. and makes it possible to find its quantitative expression in light of the common physical definition of E., given in §1.

Engels emphasizes this fact particularly clearly and vividly in the following remarks: "The indestructibility of motion cannot be conceived merely quantitatively; it must also be conceived qualitatively; matter whose purely mechanical change of place includes indeed the possibility under favourable conditions of being transformed into heat, electricity, chemical action, life, but which is not capable of producing these conditions from out of itself, such matter has *forfeited motion*; motion which has lost the capacity of being transformed into the various forms appropriate to it may indeed still have *dynamis* but no longer *energeia*, and so has become partially destroyed."[17]

Since the universal substance is matter in motion in its spontaneous transformations from one form of motion into another, and since the possibility of these transformations is an inseparable property of matter in motion, there can be no matter without energy, just as there can be no matter without motion. This position was confirmed by relativity theory, which established the connection between mass and E.

NOTES

1. The following is a translation of Б. М. Гессен, «Энергия», *Большая Советская Энциклопедия* 64 (1933): 427–41. Hessen's original references have been substituted by English translations. This translation was completed with the aid of translation software and revised by [Translator's Note; hereafter, TN].

2. Engels, *Dialectics of Nature*, 527 [TN].

3. Leibniz, "On Nature Itself, or On the Inherent Force and Actions of Created Things [1698]," 499.

4. Lenin, "Conspectus of Feuerbach's Book *Exposition, Analysis and Critique of the Philosophy of Leibnitz*," 377.

5. Ibid.

6. Ibid., 378.

7. von Helmholtz, "On the Conservation of Force," 115 [TN].
8. Ibid., 116 [TN].
9. Ibid. [TN].
10. Ibid. [TN].
11. Engels, *Dialectics of Nature*, 370n.
12. von Mayer, *Die organische Bewegung in ihrem Zusammenhange mit dem Stoffwechsel*, 6. See Caneva, *Robert Mayer and the Conservation of Energy*, 198 [TN].
13. Mayer, *Die organische Bewegung*, 6. See Caneva, *Robert Mayer and the Conservation of Energy*, 60 [TN].
14. Mach, *The Science of Mechanics*, 248 [TN].
15. Ibid., 250.
16. Kilogram-force meter [TN].
17. Engels, *Dialectics of Nature*, 333 [TN].

Entropy

B. M. Hessen

[455] THE LAW OF ENERGY CONSERVATION AND TRANSFORMATION AND E.

This law[1], as Engels emphasizes, must be understood not only in a quantitative, but also in a qualitative sense. In physics, E. is understood as the measure of that transition of various forms of energy into thermal energy, when the latter cannot be spontaneously converted back into those forms of energy from which it arose (see *Energy*, Engels's concept of E. as a measure of motion).

Energy can be converted from one form into another, while remaining constant across all transformations. [456] There is not a single form of the motion of matter which cannot be associated with the transformation of one form of energy into another. Only by considering a given form of motion (e.g. mechanical) in complete abstraction and isolation from the forms of motion with which it is associated, do we get "pure motion within the given form," dissociated from the transformation of a given form of energy (mechanical) into others. The transformation of one form of energy into another, however, occurs differently for all forms of the motion of matter. In the case of the transition of mechanical kinetic energy into potential, this transition can occur in any direction, e.g. in a pendulum (setting friction aside), the transition of kinetic energy into potential can spontaneously occur an infinite number of times in any direction (both from kinetic to potential, and vice versa). Therefore, mechanical processes are called reversible.

The situation is different with the transition of mechanical energy into heat. The pendulum left to itself eventually stops, because its mechanical energy, due to friction against the air and inside the thread on which it is suspended, turns into thermal energy. But after the mechanical energy of the pendulum has turned into thermal energy, it cannot spontaneously turn back into mechanical energy and cause the pendulum to swing. Unlike a mechanical process, we are dealing here with an irreversible process. Thus, in the transformation of one form of energy into another, we are dealing with both reversible (the transition of potential into kinetic energy, and vice versa) and irreversible processes. In an irreversible process, the amount of energy remains constant, but the energy seems to lose the capacity to spontaneously transform from one state into another. The process of transferring heat from a warmer body to a cooler body is also irreversible. After the temperatures of the bodies are equalized, it is impossible to get variations in heat transfer, i.e. reverse heat transfer, which would spontaneously heat one body by cooling the other, which previously had the same temperature as it. A cooled glass of water cannot spontaneously heat up again by cooling the surrounding air (of the same temperature), into which it released its heat. Obviously, when converting energy, one must not only take into account its quantity, but also its quality, i.e. the state which the energy is in, since not all states of energy are equally free to transform into other forms. With all transformations of energy, a certain part of it passes into a state from which it cannot spontaneously pass into another form of motion. Part of the energy is "bound, devalued."

The measure of the transition of energy into such a state from which it cannot spontaneously pass into other forms is called E.

THE LAW OF ENERGY DISSIPATION. *"PERPETUUM MOBILE"* OF THE FIRST AND SECOND KINDS

Since there are no purely isolated processes in nature and since every process is associated with the transition of one form of energy into another, with any such transition, part of the energy passes into a "bound" state, i.e. E. appears, resulting in the irreversibility of processes. [457] The law of energy conservation shows the impossibility of building an engine, the so-called *perpetuum mobile* of the first kind, which would provide energy without drawing it from somewhere, i.e. that would create energy. It also turns out that it is impossible to build an engine which would work by reducing E., the so-called *perpetuum mobile* of the second kind. It is impossible, e.g. to build an engine that would work on the thermal energy contained in the ocean, and which would draw energy from it by cooling the ocean water.

The main condition for the construction of any engine is thus, not only the constancy of the amount of energy, but also its presence in such a state that would make it possible to spontaneously transform into another form or spontaneously transition into another state, e.g. the transition from a higher thermal level to a lower one. A heat machine can only work if there are differences in thermal levels—the temperature of a boiler and a refrigerator. As soon as the differences in thermal levels equalize, the energy is in such a state that it is no longer capable of transforming spontaneously into another state; E. has reached a maximum and all motion has ceased.

The end result of the operation of heat engines, like any process of energy conversion, is an increase in E. Thus, classical thermodynamics was based on the two following propositions: the energy of the world remains constant, and E. always increases.

THE MAIN STAGES IN THE DEVELOPMENT OF THE DOCTRINE OF E. THE IDEALISTIC INTERPRETATION OF E. THE "HEAT DEATH" OF THE WORLD

Like all thermodynamics, the doctrine of E. is inextricably linked and developed along with the development of the steam engine. The development of the steam engine immediately posed the problem of the most economical use of fuel. Watt quite clearly set himself the task of "lessening the consumption of steam, and consequently fuel, in fire engines."[2] The contract that Watt and Boulton entered into with the owner of the coal mines was that they were paid a third of the amount received from saving fuel costs. The matter of the technical rationalization of the steam engine became a central problem. The implementation of this task in practice made it necessary to study in detail the physical processes taking place in the machine.

In contrast to Newcomen, Watt, at the laboratory of the Univ. of Glasgow, studied in detail the thermodynamic properties of vapors and thereby laid the foundation for thermodynamics. Thermodynamics not only received an impetus to its development from the steam engine, but actually developed on its study. The need arose not only to study individual physical processes occurring in the steam engine, but also for a general theory of the efficiency (COP[3]) of the steam engine.

In his work "*On the Motive Power of Heat*," Sadi Carnot points out that it is necessary to provide a complete theory of the heat engine, similar to that which was available for mechanical engines. To do this, you needed to study and anticipate the effects of heat on a body. He raises the question of "whether [458] the motive power of heat is unbounded [that is, whether the amount of work that can be obtained under given thermodynamic conditions is limited or not], whether the possible improvements in steam-engines have an assignable limit,—a limit which the nature of things will not allow to be passed by any means whatever; or whether, on the contrary, these improvements may be carried on indefinitely."[4] This formulation of the question led Carnot to solve the problem of the efficiency of the steam engine, which turned out to depend solely on the temperature of the boiler and refrigerator. Thus, Carnot established that the necessary condition for obtaining mechanical work by means of a steam engine is the transfer of heat from a warmer body (the boiler) to a cooler one (the refrigerator).

As we have already seen, such a transition is always associated with irreversible processes; therefore, obtaining mechanical work in a heat engine or, in other words, converting a thermal form of motion into a mechanical one, is always associated with an increase in E. Coming very close to a general formulation of the law of E., Carnot was unable to formulate this law in a general form, e.g. he considered heat not as a form of motion, but as a certain unique weightless form of matter called "caloric."

As Engels points out, he "almost got to the bottom of the question. It was not the lack of factual data that prevented him from completely solving it, but solely a preconceived *false theory*. Moreover, this false theory was not one which had been forced upon physicists by some variety of malicious philosophy, but was one concocted by the physicists themselves, by means of their own naturalistic mode of thought, allegedly so very superior to the metaphysical-philosophical mode of thought."[5]

The results of Carnot's research were analytically and mathematically formalized by Clapeyron, who gave them a form that has remained relevant to this day (see *Carnot Cycle*).

In the Carnot Cycle, work along the isotherm can be represented as follows:

$$a_1 = \int pdv = \frac{RT_1}{m} \ln \frac{v_2}{v_1}$$

Work along the adiabat can be represented as:

$$a_2 = \int pdv = -\int C_v d = C_v (T_1 - T_2)$$

The same goes for the second isotherm and the second adiabat in T_2. Taking account of the work accomplished during the entire process, we get:<

$$A = \frac{R}{m}(T_1 - T_2) \, ln \frac{v_2}{v_1}, \text{ since } \frac{v_2}{v_1} = \frac{v_3}{v_4}$$

The efficiency of a machine is equal to $\frac{A}{q_1}$, where A refers to all the mechanical work exerted by the machine, and q_1 refers to the amount of thermal energy provided by the boiler. For the Carnot Cycle with an ideal gas of $A = q_1 - q_2$, where q_1 is the amount of heat supplied to the refrigerator, the efficiency can be represented as:

$$\eta = \frac{q_1 - q_2}{q_1}$$

Since

$$q_1 = \frac{R}{m} T_1 \ln \frac{v_2}{v_1}; \; q_2 = \frac{R}{m} T_2 \ln \frac{v_3}{v_4},$$

then

$$\eta = \frac{\frac{R}{m}(T_1 - T_2)\ln \frac{v_2}{v_1}}{\frac{R}{m} T_1 \ln \frac{v_2}{v_1}} = \frac{T_1 - T_2}{T_1}.$$

Thus, it turns out that the efficiency of a heat engine solely depends on the temperature of the boiler and the refrigerator, and not on the working substance (water vapor or some other substance). Moreover, this reasoning clearly shows that the exertion of work $\int pdv$ is associated with the transfer of heat from a boiler with a temperature T_1 to a refrigerator with a temperature of T_2 ($T_1 > T_2$).

The concept of E. receives further refinement and mathematical formulation in the works [459] of W. Thomson (Kelvin) and Clausius. These works date back to the fifties and sixties of the 19th Century. Although Clausius was one of the first to attempt to consider heat as a form of motion and one of the founders of the kinetic theory of heat, in his works on E., he does not try to explain it based on the kinetic theory of heat, but only provides a mathematical description and clarification of the concept of an irreversible process based on the Carnot Cycle.

For Clausius, entropy is a function whose direction of change is characterized by the property of the asymmetry of an irreversible process. It is therefore a function wherein two consecutive values, which characterize the two states of a body (or a system of bodies) in an irreversible process, can never be equal. Based on this, Clausius in his work of 1867 makes an attempt to apply the concept of E. to the entire world, taken as a whole, and comes to two main conclusions: (1) the energy of the world is constant; (2) the entropy of the world tends to a maximum. Since E. increases in any irreversible process, and the entire world, taken as a whole, tends to such a state in which E. reaches its maximum, Clausius then draws the conclusion that the entire world, taken as a whole, tends to a final state, from which it will never be able to spontaneously return, because such a return would be associated with a decrease in E. and would therefore violate Clausius' second position.

So, according to Clausius, the world tends to an inevitable end, to so-called "heat death."

CRITICISM OF CLAUSIUS' THEORY OF E. AND "HEAT DEATH," AS GIVEN BY ENGELS, AND HIS CONCEPT OF E. BOLTZMANN'S KINETIC INTERPRETATION OF ENTROPY

Engels provides an exhaustive, principled critique of Clausius' theory. "*Clausius' second law, etc.*, however it may be formulated, shows energy as lost, qualitatively if not quantitatively. *Entropy cannot be destroyed by natural means but it can certainly be created.* The world clock has to be wound up, then it goes on running until it arrives at a state of equilibrium from which only a miracle can set it going again. The energy expended in winding has disappeared, at least qualitatively, and can only be restored by an *impulse from outside*. Hence, an impulse from outside was necessary at the beginning also, hence, the quantity of motion, or energy, existing in the universe was not always the same, hence, energy must have been created, i.e. it must be creatable, and therefore destructible. *Ad absurdum!*"[6]

A one-sided quantitative interpretation of the law of energy conservation and the increase of E. together with the absence of a connection between E. and the concept of heat as a form of motion in Clausius' concept inevitably led him to the introduction of a divine, supernatural principle that gave the initial impulse to the world.

Engels' criticism is directed primarily against a purely quantitative understanding of the law of energy conservation and against a mechanical understanding of motion. "The *quantitative* constancy of motion," Engels writes, "was already enunciated by Descartes, and indeed almost in the same words as now by? (Clausius, Robert Mayer, Maxwell?). On the other hand, the transformation of the *form* of motion was [460] only discovered in 1842 and this, not the law of quantitative constancy, is what is new."[7] Motion is inseparable from matter and indestructible. "The indestructibility of motion," Engels emphasizes, "cannot be conceived merely quantitatively; it must also be conceived qualitatively; matter whose purely mechanical change of place includes indeed the possibility under favourable conditions of being transformed into heat, electricity, chemical action, life, but which is not capable of producing these conditions from out of itself, such matter has *forfeited motion*; motion which has lost the capacity of being transformed into the various forms appropriate to it may indeed still have *dynamis* but no longer *energeia*, and so has become partially destroyed. Both, however, are unthinkable."[8]

Thus, Clausius' position in its absolute form is nothing more than the assertion of not only the possibility, but the inevitability of the transition of matter into a state in which it will be devoid of motion. Indeed, the energy of the world wherein E. has reached its maximum is no longer able to generate the conditions for further transformation into other types, since the irreversible process has reached its end. Matter with a maximum of E. is matter that has lost its motion. Clausius' conclusions are in irreconcilable contradiction with the basic position of dialectical materialism about the inseparability of motion from matter.

Engels' conception indicates the direction of the solution of the question, and indeed only this concept frees us from the need to introduce supernatural, divine intervention that brings the world out of the state of heat death. "[E] ither we must," Engels says, "[allow a miracle or] have recourse to a creator, or we are forced to the conclusion [that what happened one day can occur again,] that the incandescent raw material for the solar systems of our universe was produced in a natural way by transformations of motion which are *by nature inherent* in moving matter, and the conditions for which, therefore, must also be reproduced by matter, even if only after millions and millions of years and more or less by chance, but with the necessity that is also inherent in chance."[9]

Engels' conception was fully confirmed in the works of Boltzmann, who provided a kinetic interpretation of E. and showed the limitations of Clausius' formulation. Boltzmann proceeded from the general ideas of the kinetic theory of matter (gases). He considers heat to be a form of the motion of molecules in a body. The thermal state of a body can be characterized by its temperature. Temperature expresses the degree of the heat of a body. The concept of temperature is not related to any hypotheses regarding the structure of a body; it is an empirical, directly observable quantity (e.g., the position of the level of mercury in a thermometer). The law of energy dissipation in the formulation given above is based on the definition of temperature as a directly observable quantity.

If we turn to the microscopic, atomic structure of matter, we need to approach [461] the definition of the thermal state of a body differently. The degree of heat of a body, according to the kinetic theory of heat, is determined by the average energy of molecular motion. Not all molecules have the same velocity, and the energy is not evenly distributed among all of the molecules, but the velocity and energy are distributed between the molecules according to a certain law. Therefore, we are talking about average energy (see *Kinetic Theory*).

Each specific distribution of energy between molecules corresponds to a specific thermal state of a body. In other words, each microstate defined by the distribution of molecules corresponds to a macrostate defined by temperature. But one and the same macrostate, which is defined by temperature, corresponds not to one, but several microstates. One and the same thermal state of a body can be realized by different distributions of energy between molecules, as long as the average energy remains the same. In fact, in order to determine the average energy, the number of molecules with a given energy is important for us, but it does not matter which molecules have this or that energy. What energy this or that individual molecule possesses is a matter of chance; what is important for us is the distribution of energy that characterizes the entire set of molecules.

Thus, the same thermal state can be realized by a large number of microstates. If we calculate the ratio of the number of microstates that characterize a given thermal state, e.g., of a gas, to the total possible number of microstates that is generally possible for a given gas, we then get the so-called thermodynamic probability. The greater the thermodynamic probability, the easier this state can be realized. Thus, from the point of view of the kinetic theory of matter, each thermal state has a certain probability. If two macrostates are given, then it is easy to determine which of the two is more likely by considering them from the point of view of their atomic-molecular structure. In observing the course of thermal processes, we see that the difference between the levels of macroscopic thermal states tends to equalize, i.e. to reach equilibrium. Since we constantly observe only a certain course of such a process, we conclude that micro-processes that reach equilibrium are most probable. If we have one volume of gas which is hotter than another and we connect both volumes, it is more likely that both will reach the same temperature than that each volume remains at different temperatures. The thermal process will be such that the temperature equalizes, i.e. reaches the most probable microstate, which is the state of thermal equilibrium.

It is now clear why Clausius considered thermal processes to be irreversible. They are irreversible because the course of the thermal process is a transition from a less probable state to a more probable one. But with a statistical interpretation, irreversibility loses its absolute value. The probability of a reverse transfer of heat from a warmer body to a cooler body or a deviation from thermal equilibrium is small, but not equal to zero. Consequently, [462] Clausius' formulation that E. *always* increases is incorrect. E. increases in the vast majority of cases, but not always. Precisely because heat is the motion of molecules, such forms of their motion (microstates) where entropy decreases will inevitably be realized. Boltzmann's concept is an expression of thermal laws in the form of a statistical law.

H*-Theorem and E.** The central point in the kinetic interpretation of E. given by Boltzmann is the so-called ***H-Theorem. Its essence can be reduced to the following. Consider two groups of molecules: ***A*** and ***B***. Let ***f*** denote the probability that a molecule has a certain velocity (for example, velocity with a component along the axes ***u***, ***v***, ***w***), and let ***N*** be the number of molecules per unit volume. Then, ***f N*** = ***F*** will denote the number of molecules with a certain velocity (***F*** will be a function of ***u***, ***v***, ***w***), and the number of molecules in the $\Delta u \Delta v \Delta w$ range of group ***A*** can be shown as:

$$(1)\ F\ (uvw)\Delta u\Delta w$$

Similarly, in group ***B*** (interval Δu'Δv'Δw'), it can be shown as:

$$(2)\ F(u'v'w')\ \Delta u'\Delta v'\Delta w'$$

As proven by the kinetic theory of gases, the number of collisions between groups of molecules ***A*** and ***B*** during Δt will be:

$$(3)\ v_a \Delta t = FF^{1\delta 2c}{}_r\ cos\theta' \Delta u \Delta v \Delta w \Delta u' \Delta v' \Delta w' \Delta t$$

where δ refers to the diameter of the molecules, to the relative velocity of the colliding molecules, θ to the angle between the collision axis and the relative velocity, and Δk expresses the solid angle that determines the oscillation of the collision axis.

In a state of equilibrium, the number of molecules with certain velocities in each group should remain unchanged. Using equation (3) for the number of collisions, we can account for the increment in the number of molecules in group ***A*** for all possible collisions with group ***B***. It will be equal to:

$$(4)\ \Delta u \Delta v \Delta w \Delta t \int\!\!\int\!\!\int\!\!\int (FF' - FF')\ \delta^2\ c_r\ cos\ \theta \Delta k \Delta u' \Delta v' \Delta w',$$

while on the other hand, the increase in ***F*** over Δt equals:

$$(5)\ \frac{\delta F}{\delta t}\ \Delta t \Delta u \Delta v \Delta w.$$

Equating (4) and (5), then, we obtain:

$$(6)\ \frac{\delta F}{\delta t} = \int\!\!\int\!\!\int\!\!\int (FF' - FF')\ \delta^2\ c_r.\ cos\ \theta \Delta k \Delta u' \Delta v' \Delta w',$$

Hence, it is clear that a sufficient condition for equilibrium will mean equating the integrand (7) ***(FF' – FF')*** to zero. But, this condition may not be necessary. To prove the necessity of this condition, Boltzmann introduces the ***H*** function, which is defined as follows:

$$(8)\ H = \int\!\!\int\!\!\int F\ ln\ F\ \Delta u \Delta v \Delta w.$$

The stationary state for ***H*** will be the same as for ***F***:

$$\frac{dH}{dt} = 0.$$

We get $\frac{dH}{dt}$ by substituting the value of $\frac{\delta F}{\delta t}$ from (6) and taking into account all possible cases of collisions:

$$(9)\ \frac{dH}{dt} = \frac{1}{4} \int\int\int\int\int\int (ln\, FF' - ln\, FF').$$

$$(FF' - FF')^{\delta 2c}{}_{r}\, cos\theta \Delta k \Delta u \Delta v \Delta w \Delta u' \Delta v' \Delta w'$$

The sub-integral function is always negative, since ***FF' – FF'*** always has the opposite sign of $ln\frac{FF'}{FF'}$. Thus, condition (7) will not only be sufficient, but will also be necessary for ***H***, and hence for ***F***, to be stationary.

It can be seen from (9) that the main property of ***H*** is such that $\frac{dH}{dt} < 0$ for non-stationary processes and such that $\frac{dH}{dt} = 0$ for stationary processes. These two equations, known as Boltzmann's "***H***-Theorem," determine the value of the function ***H***. It changes in the same way as E. But since we came to it, proceeding from the consideration of gas as a kinetic system of molecules, there is a possibility of connecting ***H*** with E. to provide a kinetic interpretation of E.

The Physical Significance of *H*. We calculate ***H*** by the formula:

$$H = \int\int\int F\, ln\, F \Delta u \Delta v \Delta w,$$

where $F = vf = va^3 e^{-\beta 2c2}$ (see *Kinetic Theory of Gases*), and then obtain:

$$H = v\, ln\,(va^3) - \frac{3}{2} r$$

and finally, where after conversion:

$$-H = v \left(\frac{5}{2}\, ln\, T - ln\, p + \text{Const}\right).$$

Here, ***H*** is expressed as a function of pressure and temperature.

[463] Calculating E. under the same assumptions, we find:

$$S = \frac{R}{N} v \left(\frac{5}{2}\, ln\, T - ln\, p + \text{Const}\right),$$

and then,

$$S = \frac{-R}{N} H = -kH$$

(where $\boldsymbol{R}$ is the gas constant, and N is the number of molecules). This is the basic relationship between $\boldsymbol{H}$ and E.

E. and Thermodynamic Probability. As discovered above, the macroscopic state is defined by temperature. It is commonly called the "thermodynamic state" (macrostate). But, any state of a body considered from the molecular-kinetic point of view is determined by the position of molecules and their velocities. The state which is defined by the arrangement of molecules and their velocities is referred to as the "microstate."

The primary task is to determine which microstates correspond to the same thermodynamic state (macrostate).

Let us imagine all molecules as enumerated (***a***, ***b***, *c*, ***d*** . . .) and establish $\boldsymbol{u} + \Delta \boldsymbol{u}$, etc. as the limits within which the velocities of the molecules lie. Let us then calculate the probability of a certain macrostate. This can be done, since, as was indicated above, a very large number of microstates correspond to the same macrostate. Let us denote the probability of the η-th molecule as having a velocity within $\boldsymbol{u}_\eta + \Delta \boldsymbol{u}_\eta$, through $\boldsymbol{w}_\eta$. Then, the probability of a gas state in which N_1 molecules have a velocity within $\boldsymbol{u}_1 + \Delta \boldsymbol{u}_1 \ldots, N_2$, and within $\boldsymbol{u}_2 + \Delta \boldsymbol{u}_2$ will be ${}_{w1}N_{1w2}N_{2\ldots w\eta}N_\eta$ $(N_1 + N_2 + \ldots + N_\eta = N$ the number of molecules in a given volume). Since it does not matter which molecules have a velocity within $\boldsymbol{u}_i + \Delta \boldsymbol{u}_i \ldots$ for the macrostate, for all possible permutations within the same group, we will obtain different microstates that correlate with the same macrostate, the probability of which (w) will obviously be (see *Probability Theory*, *Compound Theory*):

$$W = {}_wN_{1w2}N_2 \ \cdots \ {}_{w\eta}N_\eta \frac{N!}{N_1!\,N_2! \ \ldots \ N_\eta!}.$$

Since a gas is in a state of equilibrium, the probability of any molecule being in any range of velocities is the same, and we can establish $\boldsymbol{w}_1 = \boldsymbol{w}_2 \ldots \boldsymbol{w}_\eta = \boldsymbol{w}_1$, and then

$$W = \frac{{}_wN_{N!}}{N_1!\,N_2! \ \ldots \ N_\eta!}.$$

This, therefore, is the probability of such a state in which N_i molecules fall in the range of speeds $\boldsymbol{u}_i + \Delta \boldsymbol{u}_i \ldots$.

Let us compare this probability with the probability of a state $\boldsymbol{W}_0$ chosen in a certain way, in which all molecules fall in the same range of velocities $\boldsymbol{u}_1 + \Delta \boldsymbol{u}_1 \ldots$ (the so-called 'normal state', which is of course arbitrarily chosen):

$$W_0 = \frac{W^N N!}{N!} = W^N.$$

Then, the relative probability which shows how many times W is more likely than W_0 will be:

$$\frac{W}{W_0} = \frac{{}_{W}N_{N!}}{N_1!N_2! \dots N_\eta!} = W \text{ thermodynamic.}$$

In contrast to mathematical probability, this ratio, which is called "thermodynamic probability" (Planck), will be a very large number. It shows how many microstates realize the same macrostate.

Transforming the therm. Factorials $N!$ and $N_i!$ included in W according to the Stirling Formula, we obtain the following logarithm:

$$(10) - \ln W \text{ term.} = \sum N_i \ln N_i + \text{Const.},$$

But N_i is equal to the number of molecules in the range of speeds $u_i + \Delta u_i$ . . ., and as indicated above, $N_i = F_i \Delta u_i \Delta v_i \Delta w_i$.

Substituting (10) and replacing it with the integral $\sum$, we obtain:

$$(11) - \ln W \text{ term.} = \int\int\int F \ln \Delta u \Delta v \Delta w;$$

comparing (10) with (8), we get:

$$(12) \ln W \text{ term.} = H;$$

And substituting the value of H for S, we get:

$$S = k \ln W \text{ term., where } k = \frac{R}{N}$$

E. is proportional to the logarithm of thermodynamic probability. This is the basic ratio of the kinetic theory of E., obtained by Boltzmann.

The statistical interpretation of the law of energy dissipation eliminates the difficulty of the question of the beginning of the world in time. Any equilibrium that comes about [464] is a death only from a limited, "earthly" point of view. From a cosmic point of view, any minimally probable formations or deviations from thermal equilibrium are possible. Heat death is the beginning of new life. The world has no beginning or end, either in time or space. This is the final conclusion of the statistical interpretation of the law of energy dissipation.

This is how Ludwig Boltzmann (1872) formulates this conclusion of the statistical theory: "[o]ne can think of the world as a mechanical system of an enormously large number of constituents, and of an immensely long period of time, so that the dimensions of that part containing our own 'fixed stars' are minute compared to the extension of the universe; and times that we call eons are likewise minute compared to such a period. Then in the universe, which is in thermal equilibrium throughout and therefore dead, there will occur here and there relatively small regions of the same size as our galaxy (we call them single worlds) which, during the relative short time of eons, fluctuate noticeably from thermal equilibrium, and indeed the state probability in such cases will be equally likely to increase or decrease. For the universe, the two directions of time are indistinguishable, just as in space there is no up or down. However, just as at a particular place on the earth's surface we call 'down' the direction toward the center of the earth, so will a living being in a particular time interval of such a single world distinguish the direction of time toward the less probable state from the opposite direction (the former toward the past, the latter toward the future). By virtue of this terminology, such small isolated regions of the universe will always find themselves 'initially' in an improbable state. This method seems to me to be the only way in which one can understand the second law—the heat death of each single world—without a unidirectional change of the entire universe from a definite initial state to a final state."[10]

It suffices to compare Boltzmann's statements with Engels' theory, cited above, to see that Boltzmann is primarily guided by the same assumptions as Engels, although Engels was not aware of Boltzmann's work, nor was Boltzmann of Engels'. E. theory, developed by Boltzmann, received experimental confirmation in the works of Smoluchowski, who showed in his analysis of Brownian motion that deviations from Clausius' formulation take place not only on a cosmic scale, but can also be observed in laboratory conditions (see *Fluctuation*). Recently, in connection with various cosmological theories, idealistic theories about the "death of the universe" are again being revived. Although these new theories (Eddington, Lemaître) proceed not only from thermal processes, but also from the consideration of complex electromagnetic radiation processes (see *Radiation*, *Cosmic Rays*), the fundamental basis of their reasoning is that of Clausius. Engels' criticism [465] and his concept of the forms of the motion of matter remain valid and provide a clear direction in the construction of a dialectical theory of radiation.

NOTES

1. The following is a translation of Б. М. Гессен, «Энтропия», *Большая Советская Энциклопедия* 64 (1933): 455–65. Hessen's original references have been substituted by English translations. This translation was completed with the aid of translation software and revised by O. Bashkina [Translator's Note; hereafter, TN].

2. J. Watt, "Steam Engines, &." (London: George Edward Eyre and William Spottiswoode, 1855 [1769]), 2.

3. Coefficient of Performance.

4. N.-L.-S. Carnot, *Reflections on the Motive Power of Heat*, ed. R. H. Thurston (New York: John Wiley and Sons, 1897 [1824]), 42. The bracketed remarks are Hessen's [TN].

5. Engels, *Dialectics of Nature*, 400.

6. Ibid., 563.

7. Ibid., 558.

8. Ibid., 333.

9. Ibid. The bracketed remarks are Hessen's [TN].

10. L. Boltzmann, *Lectures on Gas Theory*, trans. S. G. Brush (New York: Dover Publications, Inc., 1964 [1896]), 402–3 [TN].

The Ergodic Hypothesis

B. M. Hessen

[577] **ERGODIC HYPOTHESIS,**[1] an assumption in statistical mechanics that the average value of a certain quantity (for example, velocity) with respect to each molecule (the so-called spatial average) is equal to the average value of the same quantity, but not taken with respect to all of the molecules, but in relation to one molecule over a sufficiently long period of time.

Take, e.g. the volume of a gas, which is composed of *n* molecules. It is necessary to determine the average value for the speed of molecules at a given temperature because based on Maxwell's distribution law, we know that at any temperature, there are molecules in a gas with different speeds.

There are two ways to determine the average velocity: (1) determine what speed each molecule has at *one particular moment of time*. Then, multiply the value of the speed by the number of molecules that have the same speed, add all of these products and then divide the resulting sum by the total number of molecules. This will result in the so-called spatial average, since it only accounts for the total number of molecules at *a given moment in time* and not the motion of molecules and changes in *their speed over a certain period of time*; (2) track how the velocity of a single molecule changes over a certain sufficiently long period of time. Then, multiply the value of the speed by the number of time intervals when the molecule had the same velocity, add all the products and divide by the total number of time intervals over which the observation was made (e.g. a few seconds). You will get another average value [578] called the time average. Calculating the spatial average is a simpler task and, in most cases, can be carried out by using statistical methods. Calculating the temporal average is often very difficult. If we assume that the spatial average is equal to the temporal average, i.e., by introducing the E.h., we can then replace the calculation of the temporal average with the spatial average. This is exactly what happens in statistical mechanics, when instead of investigating how one state of a system passes into another over time, a statistical calculation of specially selected systems (virtual aggregates) at one moment in time is made. The E.h. is the basis of all calculations in statistical mechanics and, recently, the whole theory of Brownian motion and its experimental verification has been shown to be based on the E.h.

In classical statistical mechanics (Boltzmann, Maxwell, Gibbs), the E.h. was accepted without proof. All attempts to provide a general proof of the equality of the spatial and temporal average failed. In 1913, Rosenthal and Plancherel showed that, in its general formulation, the E.h. is incorrect, i.e. that the equality of the temporal average and the spatial average can only occur under special conditions, but that in general, these two averages cannot be equal. From the materialistic point of view regarding the objectivity of time and space as the form of the existence of matter, this result could be predicted. Indeed, in the spatial average, we replace the study of the movement of an aggregate in time with the study of its location in space at a "frozen moment of time." Naturally, in the general case, this position of the aggregate at one time displays some characteristic features of the aggregate but cannot provide a complete picture of its further movement, because no mathematical method can reduce the development in time to a simple set of states in space at a given moment. Therefore, when replacing the temporal average with the spatial average, it is necessary to determine exactly under what conditions and within what limits this replacement is possible. This question was not raised in classical statistical mechanics and therefore, a number of laws (e.g. the law of the equipartition of energy by degrees of freedom) turned out to be incorrect for low temperatures.

The E.h. can appear wherever spatial and temporal averages are concerned. So, e.g., in old theories of population statistics (Hufeland), it was believed that in, e.g., birth statistics, it was possible to replace the long-term observation of a small human group with short-term observations of a very large group. E.g. instead of studying the birthrate of a city with 10,000 inhabitants over a hundred years, it is enough to study the statistics of a country with 100 million inhabitants in a single day. These assumptions also turned out to be wrong.

NOTES

1. The following is a translation of Б. М. Гессен, «Эргодическая гипотеза», *Большая Советская Энциклопедия* 64 (1933): 577–78. This translation was completed with the aid of translation software and revised by O. Bashkina [Translator's Note].

Einstein

H. M. Müntz and B. M. Hessen

[152] **EINSTEIN,**[1] Albert (b. 1879), the greatest physicist of our time; was a professor in Zurich, Prague, Berlin and Leiden; a member of the Berlin Academy of Sciences since 1913 and over time, a member of almost all world renowned academies of science; in 1921, he was awarded the Nobel Prize. E.'s most important work belongs to two branches: quantum theory and relativity theory.

E.'s theory of light quanta (1905) is distinguished by its extreme fundamental simplicity; according to it, light with frequency ***v*** consists of particles with energy ***hv*** and momentum ***hv/c***, where ***h*** stands for Planck's constant and *c* for the speed of light. Thanks to these concepts, it is right away possible to explain, qualitatively and to a certain extent quantitatively, the photoelectric effect (the expulsion of electrons from matter by light), a number of patterns in photoluminescence, photochemistry (see *Einstein's Law*), the Compton Effect (a decrease in the frequency of x-rays dispersed by substance) and other phenomena. Thus, the wave theory of light loses its exclusive significance, which had seemed firmly established; a number of further phenomena confirmed E.'s bold hypotheses, with the help of which N. Bohr was able to e.g. theoretically deduce the exact spectrum of the atoms of hydrogen and ionized helium (1913).

It is more difficult to explain the essence of relativity theory without a large mathematical apparatus. The relativity of mechanical motion was clear to many thinkers of earlier times (for example, Aristotle, Descartes, Galileo, Kepler, Leibniz vs. Newton, Kant), but the phenomena of optics seemed to lead to the concept of absolute motion in relation to the ether. Michelson's experiment (1881), which was set up to test the speed of light under various conditions of motion, destroyed this confidence; Fitzgerald and Lorentz's auxiliary hypothesis (1892), according to [153] which a body moving with a certain absolute speed *v* contracts in the direction of motion according to the ratio,

$$\frac{1}{\sqrt{1-\frac{v^2}{c^2}}}$$

explained the aforementioned experiment, but could not satisfy the scientific thought on its own. E.'s "special" theory of relativity (1905) provided an explanation for the discrepancy of the classical theory and experience; according to it, the formulas for the transformation of movements in space and time lead to the fact that there is no, and cannot be, an absolute concept of "simultaneity" for all spatially distinct points in the world; space and time are mathematically so closely intertwined in motion that only their union in a single four-dimensional interval amounts to an in-depth analysis of physical reality; in this case, and in contrast to the classical theory, the speed of light is independent of the state of motion of the source; it is also the upper limit of the speed of physical bodies in general. Of course, a radical change in the basic views on space and time led to the transformation of all of physics in general, and, among other things, all the phenomena of electrodynamics, for which classical physics until then could not provide a proper explanation, now became the theoretical consequences of new views.

"Special" relativity theory left certain firmly entrenched contradictions of classical physics unresolved, such as the movement of the perihelion of Mercury by 43″ (approx.) in a century, which could not have been deduced from Newton's laws of gravitation. In 1913, E. provided a further generalization of special relativity theory. In special relativity theory, the four-dimensional world of space-time is metrically (pseudo-) Euclidean; however, theoretical geometry already knew the more general metric of so-called Riemannian spaces for any number of dimensions (1867); the development of tensor calculus in mathematics in recent times (especially according to Ricci) made it possible to use relatively clear symbols for any curvilinear coordinates, where Euclidean Geometry only uses rectilinear coordinates. E. filled all of these abstractions with living physical content in his "general" relativity theory; in this respect, it obtained not only an accurate explanation of the motion of Mercury, but predicted phenomena which were later justified by experiment: for instance, the distortion of light by heavy bodies and the distortion of spectral rays of sunlight. However, general relativity theory does not provide a synthesis of gravitational and electromagnetic fields. Therefore, in recent years, E. made numerous attempts to provide a synthesis of gravitational and electromagnetic phenomena in a unified field theory. So far, these attempts have not been successful and unified field theory is far from the completeness found in special relativity theory.

Among E.'s remaining, lesser known works most important are those related to physical statistics (the study of Brownian motion, 1905–1911). In 1907, E. derived a formula that accounts for the dependence of specific heat on temperature; in 1915, he theoretically predicted and experimentally confirmed, together [154] with Haas, the appearance of torque in the magnetization of iron, which provided decisive confirmation of the hypothesis of molecular currents for explaining the phenomena of permanent magnetism.

H. M. Müntz

E.'s Philosophical Views. E.'s philosophical position is inconsistent. Materialistic and dialectical elements are intertwined with Machist attitudes, which take precedent in almost all of E.'s philosophical statements. The Machist interpretation of the epistemological problems in E. played a significant role in the dissemination of the view that relativity theory is a physical confirmation of philosophical relativism, although correctly interpreted, relativity theory provides no reason for that. Despite the revolutionary significance of relativity theory with respect to views on space and time and the interrelation between space, time and matter, it is generally a continuation of the views of classical physics on the spatial and temporal continuum by synthesizing spatial and temporal intervals into a four-dimensional continuum.

Regarding the problem of the relationship between discontinuity and continuity, E. adopts the standpoint of the primacy of continuity. This explains his attitude towards the contemporary development of quantum theory and quantum mechanics and his persistent attempts to obtain quanta from unified field theory, i.e. to derive discontinuity from continuity. The one-sided prevalence of continuity is the non-dialectic element of relativity theory, which is, by its very nature, a theory of pure continuum.

In the field of politics, E. does not belong to any political party and in his political statements, adopts a liberal democratic position. He did not disguise his sympathies for the USSR and was a member of the Society of Friends of the USSR, during the Russian famine of 1921–1922, he was the chair of the German Volga Committee for Famine Relief. In recent years, E. participated in the anti-war campaign organized by Henri Barbusse, Romain Rolland and M. Gorky. In 1933, after Hitler's fascist government came to power, in protest against the persecution of Jews in Germany and against fascist terror, E. left Germany and publicly renounced his German citizenship. The French government created a special department for E. in the Collège de France. See *Relativity Theory*, *Relativism*, *Quantum Theory*.

B. M. Hessen

NOTES

1. The following is a translation of Г. М. Мюнтц and Б. М. Гессен, «Эйнштейн», *Большая Советская Энциклопедия* 63 (1933): 152–54. This translation was completed with the aid of translation software and revised by O. Bashkina.

Bibliography

PRIMARY LITERATURE

English

Hessen, B. 2009 [1931]. "The Social and Economic Roots of Newton's *Principia*." In *The Social and Economic Roots of the Scientific Revolution: Texts by Boris Hessen and Henryk Grossmann* (pp. 41–101). Edited by G. Freudenthal and P. McLaughlin and translated by P. Schimrat. Dordrecht: Springer.

———. 2019 [1927]. "Preface to Articles by A. Einstein and J.J. Thomson." Translated by S. Winkler. *Society and Politics – Boris Hessen and the Dialectics of Natural Science* 13: 87–102.

———. 2020 [1930]. "Materialist Dialectics and Modern Physics: Abstracts of the Report at the First All-Union Congress of Physicists in Odessa on 19 August 1930." Translated by S. Winkler. *Historical Materialism* 28.4: 235–41.

———. 2021 [1928]. Selections from *The Main Ideas of the Theory of Relativity*. In *Boris Hessen: Physics and Philosophy in the Soviet Union, 1927–1931; Neglected Debates on Emergence and Reduction*. Edited and translated by C. Talbot and O. Pattison. Cham: Springer Nature.

———. 2021 [1931]. "On the Question of the Causality Problem in Quantum Mechanics. Preface to the 1931 Russian Translation of Arthur Haas, *Materiewellen und Quantenmechanik* [*Wave Mechanics and the New Quantum Theory*]." In *Boris Hessen: Physics and Philosophy in the Soviet Union, 1927–1931; Neglected Debates on Emergence and Reduction*. Edited and translated by C. Talbot and O. Pattison. Cham: Springer Nature.

———. 2021. "Selected Material on the Work of Richard von Mises." In *Boris Hessen: Physics and Philosophy in the Soviet Union, 1927–1931; Neglected Debates on Emergence and Reduction*. Edited and translated by C. Talbot and O. Pattison. Cham: Springer Nature.

———. 2021 [1927]. “Marian Smoluchowski (On the 10th Anniversary of His Death).” Translated by S. Winkler and revised by A. Kojevnikov. *Science in Context* 34.1: 137–141.

———. 2021 [1928]. “Mechanical Materialism and Modern Physics.” Translated by S. Winkler and revised by A. Kojevnikov. *Science in Context* 34.1: 155–186.

———. 2022 [1936?]. *Manuscripts and Documents on the History Physics*. Edited by P. Omodeo and S. Winkler and translated by and G. Vivaldi. Venice: Verum Factum.

———. 2022 [1933]. “Energy.” Translated by S. Winkler and revised by O. Bashkina, 120–136 (See Appendix 3)

———. 2022 [1933]. “Entropy.” Translated by S. Winkler and revised by O. Bashkina, 137–150 (See Appendix 4).

———. 2022 [1933]. “The Ergodic Hypothesis.” Translated by S. Winkler and revised by O. Bashkina, 151–152 (see Appendix 5)

———. 2022 [1931]. “Ether.” Translated by S. Winkler and revised by O. Bashkina, 117–119 (see Appendix 2).

———. 2022 [1931]. “Space.” Translated by S. Winkler and revised by O. Bashkina, 114–116 (see Appendix 1).

Hessen, B., and V. Egorshin. 2021 [1927]. “The Fifth Congress of Russian Physicists.” In *Boris Hessen: Physics and Philosophy in the Soviet Union, 1927–1931; Neglected Debates on Emergence and Reduction*. Edited and translated by C. Talbot and O. Pattison. Cham: Springer Nature.

———. 2021 [1927]. “On Cde. Timiryazev’s Attitude towards Contemporary Science.” Translated by S. Winkler and revised by A. Kojevnikov. *Science in Context* 34.1: 143–153.

Hessen, B. M., and H. M. Müntz. 2022 [1933]. “Einstein.” Translated by S. Winkler and revised by O. Bashkina, 153–156 (see Appendix 6).

Russian

Гессен, Б. М. [Hessen, B. M.]. 1927. «Предисловие к статья Г. Мие «Проблема материи» [Preface to Article ‘The Problem of Matter’ by G. Mie]». *Под знаменем марксизма* [*Under the Banner of Marxism*] 1: 118–19.

———. 1929. «Идеалистические течения в современной физике и борьба с ними [Idealistic Trends in Modern Physics and the Struggle against Them]». *Молодая гвардия [The Young Guard*] 3: 58–71.

———. 1929. «К вопросу о подготовке научной смены в области теоретического естествознания [On the Question of Training Scientific Workers in the Field of Theoretical Natural Science]». *Коммунистическая революция* [*Communist Revolution*] 5: 62–67.

———. 1931. «Пространство [Space]». *Малая Советская Энциклопедия* [*Small Soviet Encyclopedia*] 6: 949–51.

———. 1931. «Эфир [Ether]». *Большая Советская Энциклопедия* [*Great Soviet Encyclopedia*] 65: 16–18.

———. 1933. «Энергия [Energy]». *Большая Советская Энциклопедия* 64: 427–41.

———. 1933. «Энтропия [Entropy]». *Большая Советская Энциклопедия* 64: 455–65.

———. 1933. «Эргодическая гипотеза [The Ergodic Hypothesis]». *Большая Советская Энциклопедия* 64: 577–78.

———. 1935. Рецензия на Г. В. Ф. Гегель Сочинения, Т. *2: Философия природы* [Review of *Writings, vol. 2: The Philosophy of Nature* by G. W. F. Hegel]. За большевистскую книгу [*On Bolshevik Literature*] 13/14: 25–29.

———. 2018 [1931] «Выступление на заседании президиума коммунистической академии [Speech at the Meeting of the Presidium of the Communist Academy (1 August 1931)]». Edited by С. Н. Корсаков [S. N. Korsakov], *Эпистемология и философия науки* [*Epistemology and Philosophy of Science*], 3: 205–10.

———. 2019. «Выступление на научной сессии философии, посвящённой 25-летию выхода в свет труда В. И. Ленина «*Материализм и эмпириокритицизм*». 22 Июня 1934 г. [Speech at the Scientific Session of Philosophy, Dedicated to the 25th Anniversary of the Publication of V. I. Lenin's Work, *Materialism and Empirio-Criticism* (22 June 1934)]». Edited by С. Н. Корсаков. *Эпистемология и философския науки* 56.1: 216–24.

Гессен, Б. М., and С. И. Луппол [S. I. Luppol]. 1928. «О кружках по изучению диалектического материализма среди молодых научных работников [On Circles for the Study of Dialectical Materialism among Young Scientists]». *Коммунистическая революция* 14: 77–83.

Гессен, Б. М., and Г. М. Мюнтц [H. M. Müntz]. 1933. «Эйнштейн [Einstein]». *Большая Советская Энциклопедия* 63: 152–54.

Гессен, Б. М., and И. И. Подволоцкий [I. I. Podvolotskiy]. 1929. «Философские корни правого оппортунизма [The Philosophical Roots of Right-Wing Opportunism]». *Под знаменем марксизма* 9: 1–29.

Гессен, Борис, et al. 1930, «О борьбе на два фронта в философии [On the Fight on Two Fronts in Philosophy]». *Под знаменем марксизма* 5: 139–49.

SECONDARY LITERATURE

English

Althusser, L. 1971 [1968]. "Lenin and Philosophy." In *Lenin and Philosophy and Other Essays*. Translated by B. Brewster. New York: Monthly Review Press.

———. 1971 [1970]. "Ideology and Ideological State Apparatuses (Notes towards an Investigation)." In *Lenin and Philosophy and Other Essays*. Translated by B. Brewster. New York: Monthly Review Press.

Bacciagaluppi, G., and A. Valentini. 2009. *Quantum Theory at the Crossroads: Reconsidering the 1927 Solvay Conference*. Cambridge: Cambridge University Press.

Bakhurst, D. 2018. "On Lenin's *Materialism and Empirio-Criticism*." *Studies in East European Thought* 70: 107–19.

Bhaskar, R. 1981. "Empiricism." In *Dictionary of the History of Science*. Edited by W. F. Bynum, E. J. Browne and R. Porter. London: The Macmillan Press Ltd.

Bochenski, J. M. 1963. *Soviet Russian Dialectical Materialism [Diamat]*. Dordrecht: D. Reidel Publishing Company, Inc.

Bogdanov, A. 2020 [1904–1906]. *Empiriomonism: Essays in Philosophy, Books 1–3*. Edited and translated by D. G. Rowley. Leiden: Brill.

Bohr, N. 1928. "The Quantum Postulate and the Recent Development of Atomic Theory." *Nature*: 580–90.

Boltzmann, L. 1964 [1896]. *Lectures on Gas Theory*. Translated by S. G. Brush. New York: Dover Publications, Inc.

Bukharin, N. 1971 [1931]. "Theory and Practice from the Standpoint of Dialectical Materialism." In *Science at the Cross Roads – Papers Presented to the International Congress of the History of Science and Technology, Held in London from June 29th to July 3rd, 1931 by Delegates of the USSR*. Edited by N. Bukharin. London: Frank Cass and Co. Ltd.

Caneva, K. L. 1993. *Robert Mayer and the Conservation of Energy*. Princeton: Princeton University Press.

Carnot, N.-L.-S. 1897 [1824]. *Reflections on the Motive Power of Heat*. Edited by R. H. Thurston. New York: John Wiley and Sons.

Chilvers, C. A. J. 2007. "The Tragedy of Comrade Hessen: Biography as Historical Discourse." In *The History and Poetics of Scientific Biography*. Edited by T. Söderqvist. Burlington: Ashgate Publishing.

Clausius, R. 1867. *The Mechanical Theory of Heat, with Its Applications to the Steam-Engine and to the Physical Properties of Bodies*. Edited and translated by T. A. Hirst. London: John van Voorst.

Cohen, H. F. 1994. *The Scientific Revolution: A Historiographical Inquiry*. Chicago: University of Chicago Press.

Deborin, A. 1929. *Hegel and Dialectical Materialism*. Translator unknown. www.marxists.org.

Deletete, R. J. 2005. "*Die Lehre von der Energie*: Georg Helm's Energetic Manifesto." *Centaurus* 47: 140–62.

Descartes, R. 1985 [1647]. *Principles of Philosophy*. In *The Philosophical Writings of Descartes, vol. 1*. Edited and translated by J. Cottingham, R. Stoothoff and D. Murdoch. Cambridge: Cambridge University Press.

Dietzgen, J. 1906. *Excursions of a Socialist in the Domain of Epistemology*, in *Some of the Philosophical Essays by Joseph Dietzgen*. Translated by M. Beer and T. Rothstein. Chicago: Charles H. Kerr.

Einstein, A. 1954 [1927]. "The Mechanics of Newton and Their Influence on the Development of Theoretical Physics." In *Ideas and Opinions*. Edited by C. Seeling, et al. and translated by S. Bargmann. New York: Crown Publishers.

———. 1955 [1949]. "Why Socialism?" *Monthly Review* 7.2: 44–50.

———. 1989 [1905]. "On the Electrodynamics of Moving Bodies." In *The Collected Papers of Albert Einstein, vol. 2: The Swiss Years – Writings, 1900–1909*. Edited by A. Beck and P. Havas. Princeton: Princeton University Press.

———. 1989 [1905]. "Does the Inertia of a Body Depend Upon Its Energy Content?" In *The Collected Papers of Albert Einstein, vol. 2: The Swiss Years – Writings, 1900–1909*. Edited by A. Beck and P. Havas. Princeton: Princeton University Press.

———. 2002 [1920]. "Ether and The Theory of Relativity." In *The Collected Papers of Albert Einstein, vol. 7: The Berlin Years – Writings, 1918–1921*. Edited by A. Engel and E. Schucking. Princeton: Princeton University Press.

Engels, F. 2010 [1880]. *Socialism: Utopian and Scientific*. In *Marx and Engels: Collected Works, vol. 24; Marx and Engels 1874–83*. Edited by Y. Kalinina and translated by D. Forgacs, et al. London: Lawrence and Wishart.

———. 2010 [1883]. *Dialectics of Nature*. In *Marx and Engels: Collected Works, vol. 25; Engels*. Edited by N. Karmanova, et al. and translated by C. Dutt. London: Lawrence and Wishart.

———. 2010 [1886]. *Ludwig Feuerbach and the End of Classical German Philosophy*. In *Marx and Engels: Collected Works, vol. 26; Engels 1882–89*. Edited by B. Tartakovsky and translated by N. Jacobs, et al. London: Lawrence and Wishart.

Freudenthal, G. 2005. "The Hessen-Grossmann Thesis: An Attempt at Rehabilitation." *Perspectives on Science* 15.2: 166–93.

Freudenthal, G., and P. McLaughlin. 2009. "Classical Marxist Historiography of Science: The Hessen-Grossmann Thesis." In *The Social and Economic Roots of the Scientific Revolution: Texts by Boris Hessen and Henryk Grossmann*. Edited by G. Freudenthal and P. McLaughlin. Dordrecht: Springer.

———. 2009. "Boris Hessen: In Lieu of a Biography." In *The Social and Economic Roots of the Scientific Revolution: Texts by Boris Hessen and Henryk Grossmann*. Edited by G. Freudenthal and P. McLaughlin. Dordrecht: Springer.

Gorelik, G., and A. W. Bouis. 2005. *The World of Andrei Sakharov: A Russian Physicist's Path to Freedom*. Oxford: Oxford University Press.

Gorelik, G. E., and V. Ya. Frenkel. 1994. *Matvei Petrovich Bronstein and Soviet Theoretical Physics in the Thirties*. Translated by V. M. Levina. Basel: Springer Basel AG.

Graham, L. R. 1985. "The Socio-Political Roots of Boris Hessen: Soviet Marxism and the History of Science." *Social Studies of Science* 15: 705–22.

———. 1987. *Science, Philosophy, and Human Behavior in the Soviet Union*. New York: Columbia University Press.

———. 1993. *Science in Russia and the Soviet Union: A Short History*. Cambridge: Cambridge University Press.

Grossmann, H. 2011. "The Social Foundations of the Mechanistic Philosophy and Manufacture." In *The Social and Economic Roots of the Scientific Revolution: Texts by Boris Hessen and Henryk Grossmann*. Edited by G. Freudenthal and P. McLaughlin and translated by G. Shalit. Dordrecht: Springer.

———. 2011. "Descartes and the Social Origins of the Mechanistic Concept of the World." In *The Social and Economic Roots of the Scientific Revolution: Texts by Boris Hessen and Henryk Grossmann*. Edited by G. Freudenthal and P. McLaughlin and translated by G. Shalit. Dordrecht: Springer.

———. 2011. "Additional Texts on Mechanism." In *The Social and Economic Roots of the Scientific Revolution: Texts by Boris Hessen and Henryk Grossmann*. Edited by G. Freudenthal and P. McLaughlin and translated by G. Shalit. Dordrecht: Springer.

Halley, E. 1960 [1686]. "Halley to Newton (29 June 1686)." In *The Correspondence of Isaac Newton, vol. 2 – 1676–1687*. Edited by H. W. Turnbull. Cambridge: Cambridge University Press.

Hedeler, W. 2018. "Lenin and Stalin, Theory and Politics." In *The Palgrave Handbook of Leninist Political Philosophy*. Edited by T. Rockmore and N. Levine. London: Palgrave Macmillan.

Hegel, G. W. F. 1991. *The Encyclopaedia Logic (with the Zusätze), Pt. 1*. Translated by T. F. Geraets, W. A. Suchting and H. S. Harris. Indianapolis: Hackett Publishing Company, Inc.

———. 2004 [1830]. *Philosophy of Nature; Encyclopaedia of the Philosophical Science, Pt. 2 (1830)*. Translated by A. V. Miller. Oxford: Clarendon Press.

Herzen, A. 1946. "Letter 1: Empiricism and Idealism." In *Selected Philosophical Works*. Translated by L. Navrozov. Moscow: Foreign Languages Publishing House.

Ienna, G., and G. Rispoli. 2019. "Boris Hessen at the Crossroads of Science and Ideology: From International Circulation to the Soviet Context." *Society and Politics – Boris Hessen and the Dialectics of Natural Science* 13: 37–64.

Ilyenkov, E. 1982. *Leninist Dialectics and the Metaphysics of Positivism*. Translator unknown. www.marxists.org.

Joravsky, D. 1955. "Soviet Views on the History of Science." *Isis* 46.1: 3–13.

———. 2009. *Soviet Marxism and Natural Science, 1917–1932*. London: Routledge.

Josephson, P. 1991. *Physics and Politics in Revolutionary Russia*. Berkeley: University of California Press.

———. 2019. "Boris Hessen as Philosopher and Polemicist: Ideology, Politics and the New Physics in the Soviet Union in the 1920s and 1930s." *Society and Politics – Boris Hessen and the Dialectics of Natural Science* 13.1: 10–36.

Kant, I. 1998 [1781]. *Critique of Pure Reason*. Edited and translated by P. Guyer and A. W. Wood. Cambridge: Cambridge University Press.

———. 2001 [1783]. *Prolegomena to Any Future Metaphysics*, 2nd ed. Translated by J.W. Ellington. Indianapolis: Hackett Publishing Company, Inc.

Kassof, B. 2005. "A Book of Socialism: Stalinist Culture and the First Edition of the *Bolshaia Sovetskaia Entsiklopediia*." *Kritika* 6.1: 55–95.

Kojevnikov, A. 2008. "The Phenomenon of Soviet Science." *Osiris* 23: 115–35.

———. 2011. "Probability, Marxism, and Quantum Mechanics." *Yearbook for European Culture of Science* 6: 211–35.

Krementsov, N. 1997. *Stalinist Science*. Princeton: Princeton University Press.

Lamb, H. 1927. "Newton's Work in Mechanics." *Nature* 119: 33–35.

Lecourt, D. 2003 [1977]. *Proletarian Science?: The Case of Lysenko*. Translated by B. Brewster. Atlantic Highlands: Humanities Press.

Leibniz, G. W. 1989 [1698]. "On Nature Itself, or on the Inherent Force and Actions of Created Things." In *Philosophical Papers and Letters*, 2nd ed. Edited and translated by L. E. Loemker. Dordrecht: Kluwer Academic Publishers.

Lenin, V. I. 1927 [1908]. *Materialism and Empirio-Criticism: Critical Notes concerning a Reactionary Philosophy*. In *Collected Works of V. I. Lenin, vol. 13*. Edited by A. Trachtenberg and translated by D. Kvitko. New York: International Publishers Co., Inc.

———. 1972 [1915]. "Conspectus of Hegel's Book *The Science of Logic*." In *V. I. Lenin: Collected Works, vol. 38; Philosophical Notebooks*. Edited by S. Smith and translated by C. Dutt. Moscow: Progress Publishers.

———. 1972 [1925]. "On the Question of Dialectics." In *V. I. Lenin: Collected Works, vol. 38; Philosophical Notebooks*. Edited by S. Smith and translated by C. Dutt. Moscow: Progress Publishers.

———. 1973 [1922]. "On the Significance of Militant Materialism." In *V.I. Lenin: Collected Works, vol. 33; August 1921–March 1923*. Edited by D. Skvirsky and translated by G. Hanna. Moscow: Progress Publishers.

———. 1974 [1914]. *Karl Marx: A Brief Biographical Sketch with an Exposition of Marxism*. In *V.I. Lenin: Collected Works, vol. 21*. Edited by S. Smith and translated by C. Dutt. Moscow: Progress Publishers.

———. 1976 [1914]. "Conspectus of Feuerbach's *Book Exposition, Analysis and Critique of the Philosophy of Leibnitz*." In *V. I. Lenin: Collected Works, vol. 38; Philosophical Notebooks*. Edited by S. Smith and translated by C. Dutt. Moscow: Progress Publishers.

———. 1977 [1894]. *The Economic Content of Narodism and the Criticism of It in Mr. Struve's Book (The Reflection of Marxism in Bourgeois Literature)*. In *V. I. Lenin: Collected Works, vol. 1; 1893–1894*. Edited and translated by the Institute of Marxism-Leninism, Central Committee of the CPSU. Moscow: Progress Publishers.

———. 1977 [1908]. *Materialism and Empirio-Criticism: Critical Notes concerning a Reactionary Philosophy*. In *V. I. Lenin: Collected Works, vol. 14; 1908*. Edited by C. Dutt and translated by A. Fineberg. Moscow: Progress Publishers.

———. 1978 [1908]. "A Letter to A. M. Gorky (25.2.1908)." In *V. I. Lenin: Collected Works, vol. 13; June 1907–April 1908*. Edited by C. Dutt and translated by B. Isaacs. Moscow: Progress Publishers.

Mach, E. 1897 [1886]. *Contributions to the Analysis of the Sensations*. Translated by C. M. Williams. Chicago: The Open Court Publishing Company.

———. 1919 [1893]. *The Science of Mechanics: A Critical and Historical Account of Its Development*, 4th ed. Translated by T. J. McCormack. Chicago: The Open Court Publishing Co.

Marquis de Laplace, P.-S. 1902 [1814]. *A Philosophical Essay on Probabilities*. Translated by F. W. Truscott and F. L. Emory. New York: John Wiley and Sons.

Marx, K. 2010 [1845]. "Theses on Feuerbach." In *Marx and Engels: Collected Works, vol. 5*. Edited by J. Cohen, et al. and translated by W. Lough. London: Lawrence and Wishart.

———. 2010 [1857–1858]. *Economic Manuscripts of 1857–58*. In *Marx and Engels: Collected Works, vol. 28; Marx 1857–61*. Edited by L. Miskievich and translated by E. Wangermann. London: Lawrence and Wishart.

———. 2010 [1859]. *A Contribution to the Critique of Political Economy*. In *Marx and Engels: Collected Works, vol. 29; Marx 1857–61*. Edited by J. Cohen, et al. and trans. V. Schmittke and Y. Sdobnikov. London: Lawrence and Wishart.

———. *Capital, Vol. 1*. In *Marx and Engels: Collected Works, vol. 35*. Edited by A. Chepurenko and translated by Y. Vashchenko and L. Zalunina. London: Lawrence and Wishart.

Marx, K., and F. Engels. 2010 [1845]. *The German Ideology*. In *Marx and Engels: Collected Works, vol. 5; Marx and Engels 1845–47*. Edited by J. Cohen, et al. and translated by W. Lough. London: Lawrence and Wishart.

———. 2010 [1848]. *Manifesto of the Communist Party*. In *Marx and Engels: Collected Works, vol. 6; Marx and Engels 1845–48*. Edited by R. Daglish, et al. and translated by J. Cohen, et al. London: Lawrence and Wishart.

Mathesius, J. 1562. *Sarepta oder Bergpostill; sampt der Jochimßthalischen kurtzen Chroniken*. Nürnberg.

Mazliak, L. 2018. "The Beginnings of the *Soviet Encyclopedia*: The Utopia and Misery of Mathematics in the Political Turmoil of the 1920s." *Centaurus* 60: 25–51.

Michelson, A. A., and E. W. Morley. 1887. "On the Relative Motion of the Earth and the Luminiferous Ether." *The American Journal of Science* 34.203: 332–45.

Morrell, J. B. 1981. "Externalism." In *Dictionary of the History of Science*. Edited by W. F. Bynum, E. J. Browne and R. Porter. London: The Macmillan Press Ltd.

———. 1981. "Internalism." In *Dictionary of the History of Science*. Edited by W. F. Bynum, E. J. Browne and R. Porter. London: The Macmillan Press Ltd.

Newton, I. 1999 [1687]. *The Principia: Mathematical Principles of Natural Philosophy*. Translated by I. B. Cohen, A. Whitman and J. Budenz. Berkeley: University of California Press.

Ortiz, E. L., and A. Pinkus. 2005. "Herman Müntz: A Mathematician's Odyssey." *The Mathematical Intelligencer* 27.1: 22–31.

Pais, A. 2005. *Subtle is the Lord: The Science and the Life of Albert Einstein*. Oxford: Oxford University Press.

Pechenkin, A. 2019 [2013]. *L. I. Mandelstam and His School in Physics*, 2nd ed. Cham: Springer Nature Switzerland.

Plato. 1997. *Republic*. In *Plato: Complete Works*. Edited by J. M. Cooper and translated by G. M. A. Grube and C. D. C. Reeve. Indianapolis: Hackett Publishing Company.

Plekhanov, G. 1940 [1897]. *The Materialist Conception of History*. New York: International Publishers.

———. 1940 [1898]. *The Role of the Individual in History*. New York: International Publishers.

Pope, A. 1902 [1730]. "Epitaph on Sir Isaac Newton." In *Every Day in the Year: A Poetical Epitome of the World's History*. Edited by J.L. Ford and M.K. Ford. New York: Dodd, Mead and Co. https://www.bartleby.com/297/154.html.

Reed, J. 2007 [1920]. *Ten Days that Shook the World*. New York: Penguin Classics.

Rowe, D. E., and R. Schulman. 2007. *Einstein on Politics: His Private Thoughts and Public Stands on Nationalism, Zionism, War, Peace, and the Bomb*. Princeton: Princeton University Press.

Schäfer, W. 1988. "Boris Hessen and the Politics of the Sociology of Science." *Thesis Eleven* 21.1: 103–16.

Shapiro, J. P. 1969. "Rehabilitation Policy under the Post Khrushchev Leadership." *Soviet Studies* 20: 490–98.

Sheehan, H. 1993 [1985]. *Marxism and the Philosophy of Science: A Critical History*. Atlantic Highlands: Humanities Press International, Inc.

Skordoulis, C. D. 2015. "Bukharin and the Social Study of Science." *Studies in East European Thought* 67: 75–89.

Stalin, J. 1954 [1929]. "A Year of Great Change: On the Occasion of the 12th Anniversary of the October Revolution." In *J. Stalin: Works, vol. 12; April 1929–June 1930*. Edited and translated by Marx-Engels-Lenin Institute. Moscow: Foreign Languages Publishing House.

———. 2013 [1938]. *Dialectical and Historical Materialism*. New York: Prism Key Press.

Steila, D. 2018. "Lenin's Philosophy in Intellectual Context." In *The Palgrave Handbook of Leninist Political Philosophy*. Edited by T. Rockmore and N. Levine and translated by L. Pasini. London: Palgrave Macmillan.

Thomson, J. J. 1927. "Newton's Work in Physics." *Nature* 119: 36–40.

Trotsky, L. 2011 [1922]. "Attention to Theory: Letter to the Editor of *Under the Banner of Marxism*." Translated by Y. Garmize and A. Gonopolskiy. *Platypus Review* 34: 1–3.

Verburgt, L. M. 2015. "'A Terrible Piece of Bad Metaphysics'?: Towards a History of Abstraction in 19th- and early 20th-Century Probability Theory, Mathematics and Logic." PhD diss., University of Amsterdam.

von Helmholtz, Hermann. 1853 [1847]. "On the Conservation of Force; a Physical Memoir [Read before the Physical Society of Berlin on the 23rd of July, 1847. Berlin. G. Reimer]." In *Scientific Memoirs, Selected from The Transactions of Foreign Academies of Science and from Foreign Journals, vol. 1*. Edited by J. Tyndall and W. Francis and translated by J. Tyndall. London: Taylor and Francis.

von Mises, R. 1957 [1928]. *Probability, Statistics and Truth*. Translated by H. Geiringer. London: George Allen and Unwin Ltd.; New York: The Macmillan Company.

———. 1987 [1938]. "Ernst Mach and the Scientific Conception of the World." In *Unified Science: The Vienna Circle Monograph Series originally edited by Otto Neurath, now in an English Edition*. Edited by B. McGuinness and translated by H. Kaal. Dordrecht: D. Reidel Publishing Company.

Watt, James. 1855 [1769]. "Steam Engines, &." London: George Edward Eyre and William Spottiswoode.

Weird, F. 2017. "Russia Set to Unveil the World's Newest Print Encyclopedia—And Its Last?" *The Christian Science Monitor*. March 23, 2017. http://www.csmonitor .com/.

Werskey, P. G. 1971. Introduction to *Science at the Cross Roads – Papers Presented to the International Congress of the History of Science and Technology, Held in London from June 29th to July 3rd, 1931 by the Delegates of the USSR.* Edited by N. Bukharin. London: Frank Cass and Co. Ltd.

Whitehead, A. N. 1923. "The First Physical Synthesis." In *Science and Civilization.* Edited by F. S. Marvin. London: Oxford University Press.

Whittaker, E. T. 1910. *A History of the Theories of Aether and Electricity, vol. 1: From the Age of Descartes to the Close of the Nineteenth Century.* London: Longmans, Green and Co.

———. 1953. *A History of the Theories of Aether and Electricity, vol. 2: The Modern Theories, 1900–1926.* London: Thomas Nelson and Sons Ltd.

Winkler, S. 2019. Introduction – "Boris Hessen and the Dialectics of Natural Science." *Society and Politics – Boris Hessen and the Dialectics of Natural Science* 13.1: 5–9.

———. 2020. "Practice and Ideology in Boris Hessen's 'The Social and Economic Roots of Newton's *Principia.*'" *Telos* 190: 29–51.

———. 2020. "The Materialist Dialectic in Boris Hessen's Newton Papers (1927 and 1931)." *Historical Materialism* 28. 4: 202–34.

———. 2021. "Individuals and Collectives in the Philosophy of Boris Hessen." *Science in Context* 34.1: 121–136.

———. 2022. "A Pantheon of Great Ideas: Boris Hessen and the History and Philosophy of Science." In *Manuscripts and Documents on the History of Physics: A Historical Materialist Textbook.* Edited by Pietro Omodeo and Sean Winkler, 35–73. Venice: Verum Factum.

Young, R. 1990. "Marxism and the History of Science." In *Companion to the History of Modern Science*. Edited by R. C. Olby, et al. London: Routledge.

Young, T. 1807. *A Course of Lectures on Natural Philosophy and the Mechanical Arts, vol. 1.* London: William Savage.

German

Avenarius, R. 1876. *Prolegomena zu einer Kritik der reinen Erfahrung* [*Prolegomena to a Critique of Pure Experience*]. Leipzig: Fues.

———. 1921. *Kritik der Reinen Erfahrung, Bände 1–2* [*Critique of Pure Experience, vols. 1–2*]. Leipzig: O.R. Reisland.

Helm, G. 1887. *Die Lehre von der Energie – Historich-Kritisch Entwickelt: Nebst Beiträgen zu einer allgemeinen Energetik* [*The Theory of Energy: A Historico-Critical Account; with Contributions on General Energetics*]. Leipzig: Verlag von Arthur Felix.

Mie, G. 1925. *Das Problem der Materie; öffentliche Antrittsrede gehalten am 26 Januar 1925* [*The Problem of Matter; Inaugural Public Address given on 26 January 1925*]. Freiburg in Baden: Speyer and Kearner.

Ostwald, W. 1902. *Vorlesungen über Naturphilosophie, vol. 2* [*Lectures on Natural Philosophy, vol. 2*]. Leipzig: Verlag von Veit and Comp.

Schrödinger, E. 1984 [1929]. "Sitzungsberichte der Preussischen Akademie der Wissenschaften, 1929 [Proceedings from the Prussian Academy of Sciences, 1929]." In *Gesammelte Abhandlungen, T. 3* [*Collected Works, vol. 3*]. Wien [Vienna]: Verlag der Österreichischen Akademie der Wissenschaften, F. Vieweg [F.Vieweg, Publisher of the Austrian Academy of Sciences].

von Mayer, J. R. 1845. *Die organische Bewegung in ihrem Zusammenhange mit dem Stoffwechsel. Ein Beitrag zur Naturkunde* [*Organic Motion in its Connection with Metabolism*]. Heilbronn: C. Drechsler.

Winkler, R. L. 2007. "Ein unveröffentlichtes Manuskript von Boris M. Hessen: 'Materialien und Dokumente zur Geschichte der Physik' (Druckfahnen – 1936 [?], ca. 700 Seiten, russisch): Kurzvortrag vor der Klasse für Sozial- und Geisteswissenschaften am 14.12.2006, gewidmet Boris Hessen anläßlich seines 70. Todestages am 20. Dezember 2006]." *Sitzungsberichte der Leibniz-Sozietät der Wissenschaften zu Berlin* 92: 133–52

Russian

Борданов, А. А. 1899. *Познание с исторической точки зрения* [*Cognition from a Historical Point of View*]. Санкт-Петербург [Saint Petersburg].

———. 1904. «Развитие жизни в природе и обществе [The Development of Life in Nature and Society]». In *Из психологии общества* [*The Psychology of Society*]. Санкт-Петербург: Электропечатня товарищества «Дело» [Electronic Printing "Press"].

Богданов, А. А., et al. 1910. *Очерки* по философии *марксизма* [*Studies in Marxist Philosophy*]. Москва [Moscow]: Звено [Link].

Ермичёв, А. А. [Ermichëv, A. A.]. 2012. Рецензия на *Журнал* Под знаменем марксизма *1922–1944. Указатель содержания* [Review of *The Journal* Under the Banner of Marxism *1922–1944. Index of Contents*]. *Вопросы философии* [*Problems of Philosophy*] 7: 184–86.

Кирсанов, В. С. [Kirsanov, V. S.]. 2010. *Избранные труды. Воспоманния коллег и друзей. Стики. Ризунски* [*Selected Works. Memories of Colleagues and Friends. Poems. Drawings*]. Edited by Д. А. Баюк [D. A. Bayuk]. Москва: Издательство им. Сабашниковых [Sabashnikov Publishing House].

Корсаков, С. Н. [Korsakov, S. N.]. 2018. «Слово товарищу гессену [The Floor is Given to Mr. Hessen]». *Эпистемология и философия науки* [*Epistemology and Philosophy of Science*] 3: 202–4.

Корсаков, С. Н., А. В. Козенко, and Г .Г. Грачева [Korsakov, S. N., A. V. Kozenko, G. G. Gracheva]. 2015. *Борис Михайлович Гессен*. 1893–1936 [*Boris Mikailovich Hessen. 1893–1936*]. Москва: Наука [Science].

Ламб, Г. [Lamb, H.]. 1927. «Работы Ньютона в области механики [Newton's Work in Mechanics]». *Под знаменем марксизма* [*Under the Banner of Marxism*] 4: 182–85.

Мещеряков, Н. Л. [Meshcheryakov, N. L.], ed. 1931. *Малая Советская Энциклопедия* [*Small Soviet Encyclopedia*] 6.

———. 1937. «Меньшевиствующий идеализм [Menshevik Idealism]». *Малая Советская Энциклопедия* 6: 822–23.

Ми, Г [Mie, G.]. 1927 [1925]. «Проблема материи [The Problem of Matter]». *Под знаменем марксизма 1*: 120–33.

Митин, М.Б., В. Н. Ральцевич, and П. Ф. Юдин [Mitin, M. B., V. N. Raltsevich and P. F. Yudin]. 1930. «О новых задачах марксистско-ленинской философии [On the New Tasks of Marxist-Leninist Philosophy]». *Правда* [*Truth*], 7 June 1930.

Столярова, О. [Stolyarova, O.]. 2017. «Идеи Б.М. Гессена и отечественная философия [The Ideas of Boris Hessen and Russian Philosophy]». *Философский журнал* [*The Philosophy Journal*] 10: 112–32.

Томсон, Дж. Дж. [Thomson, J. J.]. 1927. «Работы Ньютона в области физики [Newton's Work in Physics]». *Под знаменем марксизма* 4: 174–81.

Шмидт, О. [Schmidt, O.], ed. 1931. *Большая Советская Энциклопедия* [*Great Soviet Encyclopedia*] 65.

———. 1933. *Большая Советская Энциклопедия* 64.

———. 1933. *Большая Советская Энциклопедия* 63.

Эйнштейн, А. [Einstein, A.]. 1927. «Механика Ньютона и ее влияние на развитие теоретической физики [The Mechanics of Newton and Their Influence on the Development of Theoretical Physics]». *Под знаменем марксизма* 4: 166–73.

Index

www.ingramcontent.com/pod-product-compliance
Ingram Content Group UK Ltd.
Pitfield, Milton Keynes, MK11 3LW, UK
UKHW041838031026
15587UKWH00002B/854

9 781538 147580